Bibliografische Information der Deutschen Nationalbibliothek

Die Deutsche Nationalbibliothek verzeichnet diese Publikation in der
Deutschen Nationalbibliografie; detaillierte bibliografische Daten sind
im Internet über http://dnb.d-nb.de abrufbar.

ISBN 978-3-8325-1988-9

Logos Verlag Berlin GmbH
Comeniushof, Gubener Str. 47,
10243 Berlin
Tel.: +49 030 42 85 10 90
Fax: +49 030 42 85 10 92
INTERNET: http://www.logos-verlag.de

Rheology and Fourier-Transform Rheology on water-based systems

Dissertation
zur Erlangung des Grades

Doktor der Naturwissenschaften

am Fachbereich Chemie
der Johannes Gutenberg-Universität Mainz
vorgelegt von

Christopher Klein
geboren in Bad Kreuznach

Mainz April 2005

Erster Berichterstatter: Prof. Dr. H.W. Spiess
Zweiter Berichterstatter: Prof. Dr. H. Frey

Tag der mündlichen Prüfung: 6. April 2005

Für Meine Eltern

"Aus einem Buch abschreiben - Plagiat.
Aus zwei Büchern abschreiben - Essay.
Aus drei Büchern abschreiben - Dissertation.
Aus vier Büchern abschreiben - ein fünftes gelehrtes Buch."

Franz Molnar (1878-1952, ungar. Schriftsteller)

Contents

List of abbreviations

$(NH_4)_2S_2O_8$ APS ammonium peroxodiusulfate

$< D >$ intensity-weighted averaged diffusion coefficient

A area

$AIBN$ azo-*bis*-isobutyrylnitrile

$ARES$ Advanced Rheometer Expansion System

C stress optical coefficient

D diffusion coefficient

DLS dynamic light scattering

D_r rotational diffusion coefficient

E electric field strength

F force

FT Fourier Transformation

$G'(\omega)$ storage modulus

$G''(\omega)$ loss modulus

$G^*(\omega) = G'(\omega) + iG''(\omega)$ complex modulus

HMT Hexamethylenetetramine

$K_2S_2O_8$ KPS potassium peroxodiusulfate

K_n cumulant

$LAOS$ large amplitude oscillatory shear

LDV LASER Doppler velocimetry

$NaSS$ sodium styrene sulfonate

OAM Optical Analysis Module

$P(EO - co - AA)$ Poly(ethyleneoxide-co-acrylicacid)

PCS photon correlation spectroscopy

PDI particle size distribution index

$PEO - b - MA$ Polyethyleneoxide-b-polymethacrylicacid

$PS - DOP$ Polystyrene in Dioctyl phthalate

SDS sodium dodecyl sulfate

SEM scanning electron microscopy

T absolute temperature in K

T_g glass temperature

$Tris - Cl$ Tris (hydroxy methyl)-aminomethan chloride

UV ultraviolet spectroscopy

U_E electrophoretic mobility

$Zn(NO_3)_2$ Zinc nitrate

ZnO Zinc oxide

$\Delta n'$ birefringence

$\Delta n''$ dichroism

Θ scattering angle

δ phase lag between real and imaginary part of $G^*(\omega)$

δ' retardation

ϵ dielectric constant

η viscosity

$\eta'(\omega)$ viscous contribution of complex viscosity

$\eta''(\omega)$ elastic contribution of complex viscosity

$\eta^*(\omega)$ complex viscosity

$\frac{\omega_1}{2\pi}$ excitation frequency

γ_0 strain amplitude

λ wavelength of the light

τ correlation time

θ orientation angle

ζ zeta-potential

cmc critical micellar concentration

d optical path length

$g^{(2)}(\tau)$ correlation function

k_B Boltzmann constant

n refractive index

r_H hydrodynamic radius

t time

$tan\delta = \frac{\eta'}{\eta''} = \frac{G''}{G'}$ loss tangents

Chapter 1

Introduction

Starting from the mid 20th century polymers have become more important in industry and everyday life. For example in 1999, 190 million tons of polymers were produced worldwide [Distler 99]. Polymers are so important due to their large variety of properties, e.g. lightness, rigidity, poor conductivity, flexibility, mechanical properties, and price. In many cases, polymers are inserted as auxiliary additives, e.g. polymer dispersions. These polymer dispersions are used for protection, binding, adhesives and grafting. This diversity illustrates the technological importance of dispersions such as drilling muds, food additives, pharmaceuticals, ointments and cremes, wall paints, abrasive cleansers. They are also precursors for composite materials etc. [Larson 99]. Dispersions amount to 7 % of the produced polymers, which equals about 10 million tons of aqueous dispersions per year. The commercially most important products polystyrene-butadiene, polyvinyl-acetate, polyacrylates, and polyvinyl-ester dispersions, together make up more than 90 % of the produced dispersions. The control of the structure and flow properties of such dispersions are vital for the manufacturing process or the commercial success of the product. For example, the rheological properties of food products can often determine consumer satisfaction. In ceramic processing, dense dispersions are sometimes molded [Lange 89] and succeedingly ennobled [Rice 90, Simon 93], with the dispersions of high solid content pumped through pipes for transportation and processing. In general processes, e.g. paper coating, the dispersions can be exposed to extremely high shear rates.

The most prominent production technique for dispersions is the emulsion polymerisation [Piirma 82, El-Aasser 97], accounting for 90 % of all polymer dispersions [Distler 99]. An emulsion of monomers in water, stabilised by a surfactant, polymerises when the temperature is increased in the presence of an initiator. The energy released by the synthesis is transferred to the water bath. One of the most important properties of dispersions is the tendency to coagulate, like in the case of wall paint phase separating before use. Preventing this unintended separation by stabilisation of the particles is an important research objective. In this context the early work of Derjaguin, Landau, Vervey, and Overbeek, which resulted in the DLVO-theory, has to be mentioned. Such dispersions of small particles are often also called colloids, a term which is derived from the Greek word for glue: $\kappa\omega\lambda\lambda\alpha$. This term was introduced by Thomas Graham (1805 − 1869), defining colloids as substances that can not diffuse through a membrane. In the nineteenth century, rubber was produced from the milky sap of special trees, called latex [Larson 99]. The term latex is nowadays used for stable dispersions of polymeric particles. For "colloids" many definitions can be found. Two of the most appropriate ones will be given here. Colloids are objects of a size between atoms or molecules and macroscopic particles. Their size can range from about 1 nm over 200 nm up to $1,000$ nm. A more precise definition might be: The free enthalpy and state of particles is governed by the size of the interface. The number of molecules is not negligible compared to the number of molecules inside the particles. Colloids can be divided, according to Staudinger, in disperse colloids, molecular colloids, associated or micellar colloids, and macromolecular associates [Regitz 99].

The mechanical analysis of complex fluids via rheology is in the focus of this thesis, like the above mentioned colloidal polymer dispersions. For rheo-logical analysis several methods are available. These range from steady state over transient to dynamic measurements, which can analyse the linear and the non-linear regime. Especially the non-linear regime can give a lot of informa-tion about the structures in the examined samples. The mechanical analysis can easily enter the non-linear regime with the application of a large ampli-tude oscillatory shear strain (LAOS) [Collyer 98]. In the non-linear regime the mechanical response of rheological experiments looses the simple proportional-ity between the amplitude of the deformation and the amplitude of the stress. Therefore, a suitable simple non-linear method is needed to measure and to anal-yse the non-linear regime. By using LAOS conditions, the strain amplitude and the timescale, which induce the non-linear behaviour, can be varied independently, thus the macroscopic and also microscopic sample behaviour can be influenced and detected. By using Fourier Transformation (FT) of the stress response in combination with LAOS excitation (so-called *FT-rheology*) has recently raised interest, as a promising technique to describe non-linear phenomena of polymeric materials and dispersions [Craciun 03, Giacomin 98, Kallus 01, Krieger 73, Neidhöfer 01, Wilhelm 98, Wilhelm 00, Wilhelm 02]. The idea to analyse the mechanical non-linear regime with LAOS experiments, which then could be analysed with the Fourier-Transformation analysis was known in literature for some time [Onogi 70, Dodge 71, Krieger 73, Matsumoto 73, Davis 78, Pearson 82]. The main reasons why this technique could not be used properly before, were technical and especially computational limits. With the increase of computational power in recent years, this method is now easily available. The groups of Dealy and Giacomin used this technique within sliding-plate rheometer studies [Giacomin 98, Reimers 96, Reimers 98, Tariq 98]. By using a very sensitive detection system on commercially available rheometers, a special Fourier transformation technique and modifications on the set-up the group of Wilhelm was able to improve the signal to noise ratio by 2 to 3 decades [Wilhelm 02, Dusschoten 01].

Due to the great importance of high solid content dispersions for industry and the exposure to high shear rates in the processing, this thesis aims at the examination of water-based model systems such as colloidal dispersions under non-linear mechanical conditions. FT-rheology is a powerful method to analyse mechanical behaviour in the non-linear regime and therefore has been used as the prime method for analysis within this thesis. Preliminary examinations by Kallus et al. [Kallus 01] showed that commercially available dispersions containing spherical particles behaved highly non-linear at high solid contents. A decrease of the third harmonic intensity with increasing salt content was also found. Further investigations of polymer dispersions under LAOS conditions, performed by Carreau et al. [Craciun 03], confirmed the observation of decreasing intensities of the higher harmonics with increasing salt content. Additionally, the strain dependence of the intensity on the higher harmonics showed a maximum.

As an extension to water-based model system dispersions, aqueous dispersions of the rod-shaped particle FD-virus were also tested within this thesis. Due to its biological origin, this system is highly monodisperse and therefore a perfect model system for rod-like particles. Due to the optical transparency of the

FD-virus dispersions and the anisotropy of its refractive index, an analysis via
rheo-optical techniques was possible. Therefore, a commercially available rheo-
optical set-up was modified and extended towards FT-rheology. Tests on the FD-
virus dispersions were performed on this newly developed set-up. Additionally,
a theory developed by Dhont et al. [Dhont 03] to describe the linear and non-
linear behaviour of rod-like particles was experimentally applied to the FD-virus
dispersions.

Another subject was the examination of the influence of a defined shear field
on the mineralisation process of zinc oxide from a aqueous solution was examined.

Another important topic of this thesis was the development of an easy-to-use
method to characterise the magnitudes and phases of the higher harmonics via
known rheological phenomena. Here, not only the third and the fifth harmonic
should be explained but also the multitude of higher harmonics that can be detected in e.g. polystyrene dispersions and FD-virus dispersions. This aim was
achieved by describing the response in the time domain via a set of characteristic
functions, that correspond to the linear response, the strain softening response,
the strain hardening response and the effect of shear bands or wall slip.

In the analysis of the dispersions, special emphasis was put on the differences
of the systems synthesised via mini-emulsion polymerisation and semi-continuous
emulsion polymerisation. Here, the main focus was on the different develop-
ments of the higher harmonics magnitudes and phases of the mechanical excitation frequency. Furthermore, calculations to predict the higher harmonics were
performed.

In Chapter 2, basic rheological principles, including simple viscous and elastic
models, are reviewed. In addition, the mathematical and experimental principles
of the FT-rheology are established see Chapter 3. The emulsion polymerisation in
general and the two different reaction pathways in particular, used to synthesise
the polymer dispersions within this thesis, are introduced in Chapter 4. An introduction to the characterisation techniques of the polymer dispersions is given in
Chapter 4.2. The development of the FT-rheo-optical software and the improvements on the set-up are described in Chapter 6. The influence of polymers under
a defined shear field on the mineralisation process of ZnO is presented in Chapter
7. The experimental rheological results of dispersions are presented in Chapter

8. The development of the analysis of the time domain data via a separation in characteristic functions is shown in chapter 9. In chapter 10 the FT-rheological analysis and the analysis via the method according to Dhont of FD-virus dispersion was performed. In the last chapter, the most significant results of this thesis are summarised and an outlook for future research is given.

Chapter 2

Theory

In this chapter an introduction into rheology in general and about the mechanical properties of viscoelastic fluids in detail is given. Therefore, some phenomenological models for viscous and elastic behaviour will be presented. Additionally, the reader is familiarised to the FT-rheology technique. First, the mathematical background of the Fourier transformation is given. Later the application of FT on rheology is discussed, including the quantification of the non-linearity. Further more a short overview about the theory of non spherical particles is established.

2.1 Basic principles in Rheology

Mechanical properties like viscosity and elasticity are known for a long time. First experiments and examinations have been conducted by Newton (viscosity) and Hook (elasticity). For the purposes of this work the overview is started with Newtons equation (2.1) which is valid for steady laminar flow in a purely viscous medium:

$$\frac{F}{A} = \sigma = \eta\left(\frac{v}{d}\right) = \eta\dot{\gamma}, \tag{2.1}$$

with the force F, the area A, the shear stress σ, the viscosity η, velocity v, distance d and shear rate $\dot{\gamma}$. The case that the viscosity is independent on the shear rate is called the linear regime. When η becomes a function of $\dot{\gamma}$ the behaviour is called non-linear.

Two rheological measurement methods are introduced in the following, steady and dynamic shear. In steady shear a shear rate is applied and the measurement takes place under steady state conditions. In oscillatory shear, not only the applied strain respectively the shear rate but also the direction of shear changes permanently in a sinusoidal manner. These oscillatory measurements are called dynamic measurements due to the continuously changing shear rate. In dynamic measurements the strain amplitude γ_0 follows a sinusoidal excitation and the torque is measured. The sinusoidal movement is given by the following equation (2.2):

$$\gamma(t) = \gamma_0 sin(\omega_1 t). \tag{2.2}$$

Here γ is the deformation, γ_0 the amplitude, ω_1 the excitation frequency and t the time. When this deformation is forced on a sample, the stress will follow the excitation with a phase lag δ after a few oscillations. The resulting stress, with a phase shift δ, is given by equation (2.3):

$$\sigma(t) = \sigma_0 sin(\omega t + \delta). \tag{2.3}$$

The shear stress can be separated into two contributions, representing an in-phase and an 90 ° out-of-phase part, with respect to the excitation:

$$\sigma(t) = \sigma'(t) + \sigma''(t) = \sigma'_0 \underbrace{sin(\omega_1 t)}_{in-phase} + \sigma''_0 \underbrace{cos(\omega_1 t)}_{out-of-phase} . \tag{2.4}$$

The complex modulus equation (2.5), with the help of the Euler-relation equation (2.50), can be split up into the storage $G'(\omega)$ equation (2.6) and loss $G''(\omega)$ modulus equation (2.7):

$$G^*(\omega_1) = \frac{\sigma^*}{\gamma^*} = G' + i \cdot G'', \tag{2.5}$$

$$G'(\omega) = \frac{\sigma'_0}{\gamma_0} = \frac{\sigma_0}{\gamma_0} cos\delta, \tag{2.6}$$

$$G''(\omega) = \frac{\sigma''_0}{\gamma_0} = \frac{\sigma_0}{\gamma_0} sin\delta, \tag{2.7}$$

here the real part $G'(\omega)$ reflects the elastic and the imaginary part $G''(\omega)$ the viscous part of the complex modulus $G^*(\omega)$. Using the relation between the complex modulus $G^*(\omega)$ and the complex viscosity $\eta^*(\omega)$,

$$G^*(\omega) = i\omega\eta^*, \tag{2.8}$$

and

$$\eta^*(\omega_1) = \eta' + i \cdot \eta'', \tag{2.9}$$

and with equation (2.5) the absolute value of the complex viscosity η^* is given by:

$$|\eta^*(\omega)| = \frac{|G^*|}{\omega} = \frac{\sqrt{G'^2 + G''^2}}{\omega}. \tag{2.10}$$

Here the viscous part $\eta'(\omega)$ and the elastic contribution $\eta''(\omega)$ follow the relations:

$$\eta'(\omega) = \frac{G''}{\omega}, \tag{2.11}$$

$$\eta''(\omega) = \frac{G'}{\omega}. \tag{2.12}$$

The ratio of $\frac{G''}{G'}$ or $\frac{\eta'}{\eta''}$ defines the angle between the excitation and response is accessible and is known as the loss tangents or plainly $tan\delta$:

$$tan\delta = \frac{G''(\omega)}{G'(\omega)} = \frac{\eta'(\omega)}{\eta''(\omega)}. \tag{2.13}$$

2.1.1 Phenomenological models based on spring and dashpot

For a better understanding of the viscoelastic properties some basic phenomenological models will be introduced here. Due to the importance of oscillatory shear in this work, the focus is on the application of a periodical sinusoidal excitation. The time dependent shear strain is given by:

$$\gamma = \gamma_0 sin(\omega_1 t). \tag{2.14}$$

In the following the buildup blocks of phenomenological models to represent elasticity and viscosity, the Hookeian spring and the Newtonian dashpot will be explained. To describe more complex behaviour than just pure elastic or pure viscous, these two elements can be combined in several ways. The combination is done in such a way that the newly created models describe the behaviour in better agreement with the experimental data. Within this chapter only the simple parallel and the simple serial contribution will be discussed. Much more complicated combinations are available. The spring, based on the Hookeian law, has an ideal elastic response. With the stress $\sigma = \frac{F}{A}$ the linear dependence on the deformation γ is:

$$\sigma = G \cdot \gamma, \tag{2.15}$$

with the force F and the area A. The shear modulus G is the proportionality constant. As visible in Fig. 2.1, the stress response is in phase with the excitation,

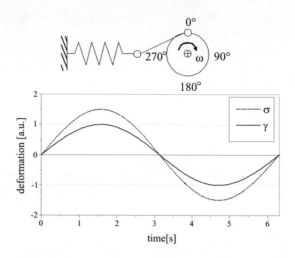

FIG. 2.1: The spring element with the excitation γ and its ideal elastic or Hookeian response σ.

meaning the phase shift δ is 0 °. The time dependent strain depends on the strain amplitude γ_0 and the angular velocity ω equation (2.2) and the shear stress is given by equation (2.3). The next simple element, corresponding to the ideal viscous Newtonian behaviour equation (2.1) is the dashpot. By exchanging the spring in Fig. 2.1 with a dashpot in Fig. 2.2 the phase difference between the excitation and the response is given by δ. This model is visualised by a piston moving in a cylinder surrounded by a lubricant. The linear dependence between stress and shear rate is called Newtonian viscosity according to Newtons law with the dynamic viscosity η as proportionality constant. The shear rate dependence is now given by:

$$\dot{\gamma} = \gamma_0 \omega_1 cos(\omega_1 t). \tag{2.16}$$

Including the equation (2.16) in equation (2.1) leads to:

$$\sigma = \eta \gamma_0 \omega_1 cos(\omega_1 t). \tag{2.17}$$

The phase difference of $\frac{\pi}{2}$ or 90 ° between the shear strain and the shear stress can be illustrated to using the relation $cos(\omega t) = sin(\omega t + \frac{\pi}{2})$:

$$\sigma = \eta \dot{\gamma}_0 \omega_1 sin(\omega_1 t + \frac{\pi}{2}). \tag{2.18}$$

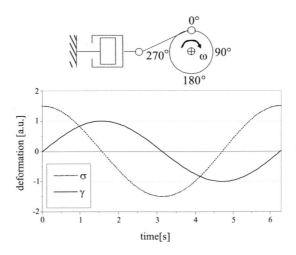

FIG. 2.2: The dashpot element with the excitation γ and its ideal viscous or Newtonian response σ.

From these results it is easy to see that in a purely viscous response strain and stress are out of phase ($\delta = 90$ °), and in a purely elastic response strain and stress are in phase ($\delta = 0$ °). It should be stated that the term linear for the shear modulus G and the dynamic viscosity η means that the stress dependence has no terms of higher order in γ, respectively in $\dot{\gamma}$. More complex systems, which denote systems where the phase can range from 0 ° i δ i 90 °, are called viscoelastic materials. The viscoelastic behaviour of these systems is described in a model by combining the two basis elements spring and dashpot. The simplest combinations of the spring and the dashpot are the Kelvin-Voigt and the Maxwell model. By arranging one dashpot and one spring parallel the resulting model, the Kelvin-Voigt model (see Fig. 2.3), describes a solid with additional viscous properties. The two different stresses, σ_d from the dashpot and σ_s from the spring, are just summed up. With the same strain for dashpot and spring $\gamma = \gamma_d = \gamma_s$ the overall stress is given by:

$$\sigma_{Kelvin-Voigt} = \sigma_d + \sigma_s = \eta \cdot \dot{\gamma} + G \cdot \gamma. \qquad (2.19)$$

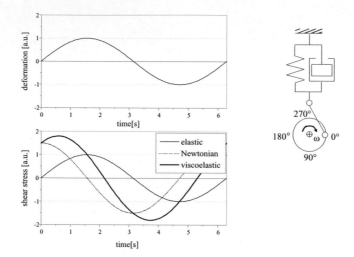

FIG. 2.3: The Kelvin-Voigt element with mainly elastic response including some viscous properties.

For γ and $\dot{\gamma}$ the above mentioned relations equation (2.14) and equation (2.16) are inserted, giving the time dependent stress of the Kelvin-Voigt model to:

$$\sigma_{Kelvin-Voigt}(t) = \eta \cdot \gamma_0 \omega_1 cos(\omega_1 t) + G \cdot \gamma_0 sin(\omega_1 t). \qquad (2.20)$$

The connection of the dashpot and the spring in series gives the description of a viscoelastic fluid. This model is called the Maxwell model (Fig. 2.4). Here the overall stress equals the stresses of the dashpot or of the spring $\sigma_{Maxwell} = \sigma_d = \sigma_s$. The overall shear rate can be calculated by rearranging equation (2.15) and equation (2.1) as follows:

$$\gamma = \frac{\sigma}{G}, \qquad (2.21)$$

$$\dot{\gamma} = \frac{d\gamma}{dt} = \frac{\sigma}{\eta}, \qquad (2.22)$$

$$\frac{d\gamma}{dt} = \frac{1}{G} \cdot \frac{d\sigma}{dt} + \frac{\sigma}{\eta}. \qquad (2.23)$$

Including equation (2.16) in equation (2.21) give an differential equation with a solution for σ:

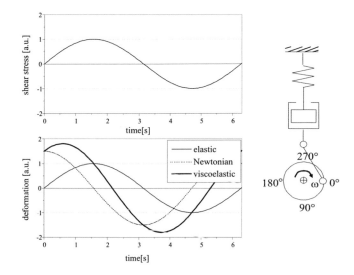

FIG. 2.4: The Maxwell element with mainly viscous response including some elastic properties.

$$\sigma = \left[\frac{G \cdot \tau^2 \cdot \omega_1^2}{1 + \tau^2 \cdot \omega_1^2} \right] \cdot sin(\omega_1 t) + \left[\frac{G \cdot \tau \cdot \omega_1}{1 + \tau^2 \cdot \omega_1^2} \right] \cdot cos(\omega_1 t), \qquad (2.24)$$

$$\sigma = G' \cdot sin(\omega_1 t) + G'' \cdot cos(\omega_1 t). \qquad (2.25)$$

The relaxation time of the system τ is defined as : $\tau = \frac{\eta}{G}$. The first term in equation (2.24) describes the elastic part G' and the second term the viscous part G''. The frequency dependence of $G' \propto \omega^2$ and $G'' \propto \omega^1$ can be used to describe the overall behaviour of a viscoelastic fluid. The behaviour of G' and G'' can be described by multiple modes for different relaxation times (see Fig. 2.5). These modes are given:

$$G'(\omega_1) = \sum_{n=1}^{n} G_n \frac{\omega_1^2 \tau_n^2}{1 + \omega_1^2 \tau_n^2}, \qquad (2.26)$$

$$G''(\omega_1) = \sum_{n=1}^{n} G_n \frac{\omega_1 \tau_n}{1 + \omega_1^2 \tau_n^2}. \qquad (2.27)$$

In this model the overall deformation amplitude γ is equal to the deformation of the individual modes. The overall shear stress $\sigma_{multimode}$ is given by $\sigma_{multimode} = \sum_{k=1}^{N} \sigma_n$. The overall stress results then in:

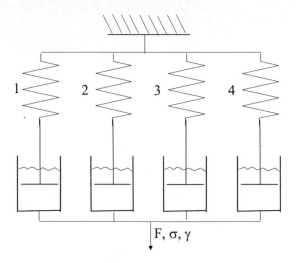

FIG. 2.5: The Maxwell multimode model consisting of several Maxwell models in parallel arrangement.

$$\sigma = \int_{-\infty}^{t} \sum_{n=1}^{n} G_n e^{\frac{-t}{\tau_n}} \dot{\gamma}_{t'} dt'.$$ (2.28)

For the description more complex experimental data of the Maxwell and the Kelvin-Voigt model is not sufficient enough to describe more complex models have been developed. The last one to mention here is the Burger model, where a Kelvin-Voigt and a Maxwell model are arranged in series. The differential equation for the stress, with the spring moduli G_1, G_2 and the dashpot viscosities η_1 and η_2, is given by:

$$\sigma = G_2 \cdot \gamma + \eta_2 \cdot \frac{d\gamma}{dt} - \eta_2 \cdot \left[\frac{1}{G_1} \cdot \frac{d\sigma}{dt} + \frac{\sigma}{\eta_1} \right].$$ (2.29)

A solution of equation (2.29) for the time dependent strain with the delay time $\tau_2 = \frac{\eta_2}{G_2}$ is shown here:

$$\gamma = t \cdot \frac{\sigma_0}{\eta_1} + \frac{\sigma_0}{G_1} + \frac{\sigma_0}{G_2} \left[1 - e^{-\frac{t}{\tau_2}} \right].$$ (2.30)

For more complex sample behaviour linear or parallel combinations of several Kelvin-Voigt or Maxwell models are used, e.g. the Burger model.

2.1.2 The Péclet number

The phenomenon of shear thinning, defined as an decrease of the viscosity depending on an increase of the shear rate, often occurs in the cases where a dispersion has a sample loading higher than 30 $vol.\%$. At small enough shear rates, the rate of regaining the equilibrium of the particles which is controlled by diffusion of the particles, is faster than the applied shear. In dilute solutions the particle diffusivity is given by:

$$D_0 = \frac{k_B T}{6\pi\eta_s a},$$ (2.31)

with η_s the solvents viscosity and a the radius of the particle. The time t_d, the particle needs to diffuse the distance of its own radius is given by:

$$t_d \approx \frac{a^2}{D_0} = \frac{6\pi\eta_s a^3}{k_B T}.$$ (2.32)

A dimensionless quantity can consequently be defined as:

$$P_e = \dot{\gamma}t_d = \frac{6\pi\eta_s\dot{\gamma}a^3}{k_B T},$$ (2.33)

and is known as the Péclet number P_e. Another definition takes the above mentioned value divided by 6π resulting in:

$$P_e = t_d\dot{\gamma} = \frac{\eta_s\dot{\gamma}a^3}{k_B T}\dot{\gamma}.$$ (2.34)

2.1.3 Pipkin diagram

In this chapter the Pipkin diagram is introduced [Pipkin 72]. Therefore the Deborah number D_e [Macosko 94] has to be introduced first. This number characterises a material in the broad range of purely viscous behaviour of Newtonian fluids, over viscoelastic behaviour to the purely elastic behaviour of rubber-like materials. It is defined as the ratio of the characteristic relaxation time λ and the characteristic flow time t_f of the material, when shear is applied. The dimensionless number is presented in equation (2.35):

$$D_e = \frac{\lambda}{t_f}.$$ (2.35)

Under oscillatory shear it is assumed that the characteristic flow time is the inverse of the angular velocity ω. Therefore $D_e = \frac{\tau}{\omega}$ is independent of the strain

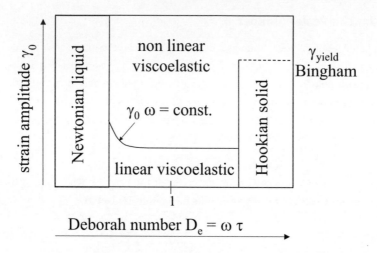

FIG. 2.6: The Pipkin diagram presenting the amount of viscoelastic behaviour of materials as a function of the strain amplitude γ_0 and the Deborah number D_e [Macosko 94].

amplitude. For the different types of polymers, their behaviour can be read from the Pipkin diagram depending on D_e. Three different characteristic types are to be mentioned:

For small Deborah numbers ($D_e \ll 1$) viscous response,

for medium sized Deborah numbers ($D_e \approx 1$) viscoelastic response,

for big Deborah numbers ($D_e \gg 1$) elastic response.

In the Pipkin diagram Fig. 2.6 the influence of the strain amplitude γ_0 and the Deborah number on the mechanical behaviour of materials under shear is visualised [Pipkin 72]. The Deborah number is plotted on the abscissa, whereas the strain amplitude is plotted on the ordinate. When the relaxation times are small compared to the flow time ($D_e \ll 1$), the polymers show a behaviour like a Newtonian liquid. In this case intermolecular processes are fast compared to the deformation time, that the energy is directly dissipated leading to a viscous behaviour. Networks are disengaged due to the high speed of the disentanglement. Furthermore the size of the strain amplitude does not influence this viscous behaviour. In the second case where the relaxation and deformation times are of the same order of magnitude the polymers show viscoelastic behaviour. The processes of building up entanglements and the disengagement of entanglements coexist. Finally, when the relaxation times are much longer than the external deformation ($D_e \gg 1$), the sample behaves like a Hookeian elastic solid. In this case the network relaxation is so slow that no entanglements are dissolved. For most samples however, like polymers, metals, and colloidal dispersion, this behaviour changes when a yield strain is reached. If the strain amplitude gets bigger than the yield strain the sample starts to flow. A further important effect is the dependence on the strain amplitude γ_0. At small γ_0 a linear viscoelastic behaviour is assumed, whereas at big γ_0 the samples show increasingly non-linear effects. Under these conditions shear strain and shear stress lose their linear dependence. Many industrial processes are conducted in this non-linear regime. Therefore methods, which are able to analyse this regime, have been developed. In the step-shear experiments the measurements are conducted parallel to the ordinate of the Pipkin diagram at constant D_e, where the characteristic time is observed by measuring the ramp time of the step. The most important experiment from our point of view is the application of a pure sinusoidal excitation with constant strain amplitude γ_0 while the frequency $\frac{\omega_1}{2\pi}$ is varied. Both experiments can be used together with the FT-rheology, especially the latter one was used for the experiments presented within this thesis.

2.2 Electrostatic interactions in colloidal systems

In colloidal systems, electrostatic interactions play an important role in the stabilisation processes of these systems. Therefore the Debye-length describing the screening effect of ions in the system, and the basics of the DLVO-theory will be introduced here.

2.2.1 Debye-length

The Debye-length gives information about the decay of the electrostatic potential, which is created by ions in e.g. a polymer dispersions. A charge in a electrolyte will exert attractive and repulsive interactions towards the surrounding ions and counter ions. These surrounding charges screen the electrostatic potential of the central-ion. The Debye-length is the characteristic length scale, that describes the screening distance of the central ions potential by the surrounding. The higher the concentration of ions in the dispersion is, the stronger is the screening effect of the ions, and therefore the shorter is the Debye-length. The Debye-length can be calculated by [Larson 99]:

$$\frac{1}{\lambda_D} = k = \left(\frac{\sum_i \rho_{i\infty} e^2 z_i^2}{\epsilon \epsilon_0 kT} \right)^{\frac{1}{2}} m^{-1}, \tag{2.36}$$

with $\rho_{i\infty}$ ionic concentration in the bulk, e the elementary charge, z_i the number of charges per ion, ϵ_0 the dielectric permittivity of vacuum, ϵ_r relative dielectric permittivity, k_B the Boltzmann constant, and T the absolute temperature in K.

2.2.2   The DLVO-theory

A short introduction to the DLVO-theory is given here. The DLVO theory describes the interactions between the particles in a typical dispersion and can therefore make predictions on the stability of the dispersions. The theory is a mathematical description for the stability of lyophobic colloids, developed by Derjagin and Landau [Derjaguin 41] and Verwey and Overbeek [Verwey 48]. Positively or negatively charged ions attach themselves to colloids and create a partly fixed and partly diffuse double layer around the particles. The centrosymmetric potential around the particles is responsible for the repulsive properties of colloids. The total particle interaction can be seen as a sum of attractive and repulsive forces. Repulsive forces have a negative sign and attractive ones have a positive sign. Both forces dependent on the radius. The energy according to the DLVO-theory is given by the sum of the potentials:

$$E_{colloids} = E_{attractive} + E_{repulsive}. \tag{2.37}$$

The attractive potential is based on the London attractive potential between two polarisable atoms:

$$E_{attractive} = -\frac{3\alpha^2 h\nu_0}{4d_p^6}, \tag{2.38}$$

with d_p the distance of the particles, α the polarisability, h the Planck's constant and ν_0 the limiting frequency of the excitation in the UV-spectra of the atom. This potential is then adapted to extended material objects like two parallel plates at a distance of d_A. Under the assumption, that the thickness of the plates is bigger than the distance d_A, it results in:

$$E_{attractive} = -\frac{A_H}{48\pi d_A^2}, \tag{2.39}$$

where A_H, the Hamaker constant $= \frac{3}{4}\pi^2 q^2\alpha^2 h\nu_0$, is summing up several prefactors. Here q is the number of surface atoms per cm^2. For the attraction of two spheres the following expression is calculated [Dörfler 02]:

$$E_{attractive} = -\frac{A_H}{6}\left[\frac{2r^2}{R^2 - 4r^2} + \frac{2r^2}{R^2} + ln\frac{R^2 - 4r^2}{R^2}\right], \tag{2.40}$$

with R the distance between the middle points of the spheres. The distance of the particles and the distance of the middle points of the spheres follow: $d_A =$

$(R - 2r)$. For small distances $(R - 2r)$ equation (2.40) simplifies to:

$$E_{attractive} = -\frac{A_H r}{12(R - 2r)}.$$ (2.41)

For big distances $(R - 2r)$ equation (2.40) simplifies to:

$$E_{attractive} = -\frac{2r^2 A_H}{3R^2}.$$ (2.42)

Note that $E_{attractive}$ depends only on R^{-2} in equation (2.42). The calculation for the electrostatic repulsion of two colloidal particles can either use the energy or the force as a function of the distance. The difficulty is that the double layers are diffuse and therefore an overlay occurs in some areas of the double layer. Particles can attract each other only when they come close to the diffuse double layer. For larger distances the particle charge is completely compensated.

The repulsive potential is defined as the difference of the free enthalpy at the distance of the particles $R - 2r$ and the free enthalpy at ∞:

$$E_{repulsive}(R - 2r) = (G - G_\infty).$$ (2.43)

The free enthalpy of the double layers can be calculated via [Dörfler 02]:

$$G = -\int_0^{\Phi_0} Q d\Phi,$$ (2.44)

with the charge Q, and the potential Φ. To get a relation between the charge Q, the potential Φ, and the distance $(R - 2r)$ the Poisson equation is used:

$$\Delta\Phi = \frac{\partial^2 \Phi}{\partial x^2} = \frac{4\pi \bar{q}_i}{\epsilon_{doublelayer}} = $$
$$\frac{4\pi e}{\epsilon_{doublelayer}}[n_- z_- e^{\frac{z_- e\Phi}{kT}} - n_+ z_+ e^{\frac{z_+ e\Phi}{kT}}].$$ (2.45)

The two double layers are seen as a capacitor and two integrations of the Poisson equation in the distance $(R - 2r)$ are done. In the next step the geometry is changed from the two parallel capacitors to a spherical one. This is achieved by splitting up the spheres into rings. Afterwards the overall repulsion is gained by the integration over all rings resulting in:

$$E_{repulsive}(R - 2r) = \frac{\epsilon r \Phi_0}{2} ln \left(1 + e^{-\kappa(R-2r)}\right),$$ (2.46)

with κ the inverse of the thickness of the double layer given by $\kappa = \frac{8\pi n z^2 e^2}{(\epsilon k_B T)^{\frac{1}{2}}}$. Here ϵ is the electric constant of the double layer, z the ion charge, e the elementary charge, and k_B the Boltzmann constant. The resulting energy of the repulsion and attraction is calculated by: $E_{colloids} = E_{attractive} + E_{repulsive}$. The parameters A_H (Hamaker constant), Φ_0 (the potential of the double layer), Q (the overall charge), and $\kappa = \frac{8\pi n z^2 e^2}{(\epsilon k T)^{\frac{1}{2}}}$ are the so-called barrier factors for stability [Derjaguin 41, Verwey 48].

2.3   Theory and practical aspects of FT-rheology

2.3.1   Fourier Transformation

Due to the general content of the Fourier transformation the following theory was rephrased from the literature [Wilhelm 02]. The data acquired in the experiments are analysed with Fourier Transformation (FT) [Bracewell 86, Ramirez 85, Wilhelm 99]. Due to the fact that the constitution of the analysed material is not changed during the experiment it is not necessary to apply more complicated analysis methods, such like the wavelet transformations [Fearn 99, Honerkamp 94]. The acquired time dependent signal is described as continually integratable function that contains periodic contributions. The FT is able to unravel these periodic contributions and assigns each of them an amplitude and phase which are frequency dependent. Instead of amplitude and phase the result can alternatively be displayed as real and imaginary part. For the analysis of the experimental data a half sided, discrete, and complex, FT was chosen. In the following, the basic mathematical principles behind the FT are introduced [Bracewell 86, Ramirez 85, Wilhelm 99]. The FT of a time domain signal s(t) or the frequency domain spectrum $S(\omega)$ is defined as:

$$S(\omega) = \int_{-\infty}^{\infty} s(t)e^{-i\omega t}dt, \qquad (2.47)$$

$$s(t) = \frac{1}{2\pi} \int_{-\infty}^{\infty} S(\omega)e^{+i\omega t}d\omega, \qquad (2.48)$$

where the prefactors may vary due to the applied convention. Generally the FT is invertible, linear, and complex over the infinite integral from $-\infty$ to $+\infty$. The basic mathematical idea behind equation (2.47) is as follows: a set of functions, e.g. polynomial, Hermite-, Laguerre- and Legendre-polynomial or harmonic functions can span, in close similarity to vectors, a space where the different functions act basically as orthogonal vectors [Wilhelm 99]. The class of oscillating functions is orthogonal with respect to all different frequencies. This space therefore has an innumerable, infinite dimension, when the infinite interval is considered. Any function (vector) $s(t)$ can now be analysed (projected) towards the specific harmonic content via the systematic projection of the different frequencies. The time dependent signal $s(t)$ is sorted by the FT with respect to frequencies $\frac{\omega}{2\pi}$ with their corresponding amplitudes and phases in a spectrum $S(\omega)$. Furthermore, it is important that any superposition of different signals in the time domain will also be a superposition in the frequency domain. The Fourier transformation is a linear transformation:

$$s(t) + g(t) \xleftrightarrow{FT} S(\omega) + G(\omega). \tag{2.49}$$

Due to the fact that the FT is inherently complex, even a real time-domain data set will become a complex frequency-domain data set with a real part $\Re(\omega)$ and an imaginary part $\Im(\omega)$. This spectra alternatively can be presented in magnitude $m(\omega)$ and phase $\Phi(\omega)$ with the relations $tan\delta = \frac{\Im}{\Re}$ and $m = (\Re^2 + \Im^2)^{\frac{1}{2}}$. The FT results can also be displayed by a cosine and sine notation due to the Euler relation:

$$e^{i\omega} = cos\omega + isin\omega. \tag{2.50}$$

When the integral in equation (2.48) is calculated from $t = 0$ to $+\infty$ it is called a half sided integral. This half-sided Fourier transformation is mostly used for experimental data analysis. In that case the Fourier transformation is very similar to a complex Laplace transformation. The time signal is not acquired continuously but in discrete time steps N and afterwards digitised. Every data point is acquired over a fixed increment t_{dw} also known as dwell time or inverse sampling rate. The overall acquisition time is given by $t_{aq} = t_{dw} \cdot N$. In Fig. 2.7 the time- and frequency-domain data, the dwell-time t_{dw}, and the acquisition-time t_{aq} of an experiment are presented. The frequency spectrum is now calculated via the Fourier transformation with N complex data points. The spectral width, also called the Nyquist frequency, is the maximum detectable frequency ν (Note, that ν has a different meaning than in chapter 2.2.2). It is given by the sampling rate $\frac{\omega_{max}}{2\pi} = \nu_{max} = \frac{1}{2t_{dw}}$. The spectral resolution is the difference between two consecutive data points $\Delta\nu = \frac{1}{t_{aq}}$. Depending on the highest frequency one wants to detect in FT-rheology [Wilhelm 99], the sampling rate should be adjusted. To be able to measure at an excitation frequency of $1\ Hz$ up to $50\ Hz$ a sampling rate of $2 \cdot 50\ \frac{1}{s}$ respectively $10\ ms$ is required. When considering the sampling rate one should bear in mind that this is calculated after the oversampling is applied see 3.1. By definition peaks in a Fourier analysis are infinitely narrow. In real measurements they have a certain width, caused by experimental inadequacies from the transducer and the motor and by the acquisition time t_{aq}. By increasing the acquisition time the actual line width is decreased and the signal

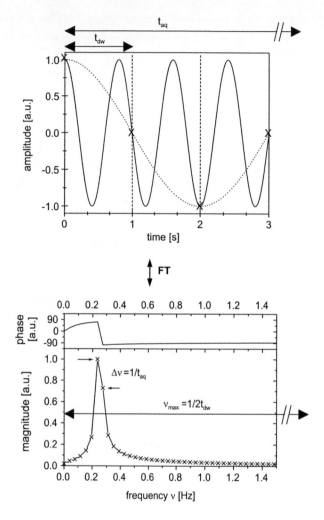

FIG. 2.7: In this schematic picture the time domain signal and its corresponding frequency domain signal are shown. The FT is done via a FFT-algorithm. Due to the discretisation of the signal and the incremental dwell time t_{dw} the detectable frequencies are ambiguous, resulting in a specific spectral width ν_{max}, which is the biggest unambiguously detectable frequency. The maximal data acquisition time (t_{aq}) limits the minimum resolution $\Delta\nu$ in the spectrum [Wilhelm 02].

to noise ratio ($\frac{S}{N}$) increased. More detailed information can be found in the literature [Bracewell 86, Claridge 99, Ramirez 85, Schmidt-Rohr 94]. In our case the signal to noise ratio is defined as the ratio of the peak-magnitude of the most intensive peak (normally at the excitation frequency, because the DC-component is ignored) divided by the standard deviation of the noise level (measured in a part of the spectra without any peak signals). To get a sufficient $\frac{S}{N}$ ratio for each measurement about 20 to 50 cycles [Wilhelm 99] are recorded leading to an overall size of $1,000$ up to $10,000$ data points N. By comparing the integral over $S(\omega)$ with $s(t)$ at $t = 0$ in equation (2.48), it is clear that $s(0)$ cannot change as a function of t_{aq} and therefore the integral over the spectrum cannot either, since the exponential term equals one for $t = 0$:

$$s(0) = \frac{1}{2\pi} \int_{-\infty}^{\infty} S(\omega) \underbrace{e^{+i\omega 0}}_{1} d\omega = \frac{1}{2\pi} \int_{-\infty}^{\infty} S(\omega) d\omega. \tag{2.51}$$

The advantage of the forced oscillation is, that it can last as it is necessary. Therefore it is not of primary interest to increase the spectral resolution by zero filling or decrease the actual line width (respectively increase the $\frac{S}{N}$ ratio) by increasing the t_{aq}. In experiments there are two possibilities to increase the $\frac{S}{N}$ ratio. First like in other FT-techniques averaging of multiple spectra can be applied and second oversampling of data points. With these methods principally an unlimited $\frac{S}{N}$ ratio can be achieved. The $\frac{S}{N}$ ratio of the averaged spectra are proportional to the square root of n spectra averaged, i. e., $\frac{S}{N} \propto \sqrt{n}$. This method is able to reduce the statistical noise. The precision of the relative intensity is increased by this method, but the phase angle information of the harmonic contributions is lost. The time domain data is shifted with respect to the excitation, so that the phase behaviour can be analysed after the FT.

The calculation of a FT used to be very computer-time consuming. The non optimised standard routine is the brute force discrete FT. It requires N^2 calculations. Therefore, several algorithms have been developed to fasten this procedure. The very common and particularly fast algorithm for discrete Fourier transformations is the FFT butterfly algorithm which only needs $N \, log_2 N$ calculations. A drawback of the simplest and most common FFT is its requirement of $N = 2^n$ data points and not arbitrary numbers like in the discrete FT [Cooley 65, Higgins 76]. This results in fixed values for the acquisition time t_{aq} and respectively for the spectral resolution $\Delta \nu = \frac{1}{t_{aq}}$. Furthermore it rarely leads, for the frequencies at

$\nu_n = n \cdot \frac{\omega_1}{2\pi}$, to be located exclusively at data points with the precise frequency or multiples of the excitation frequency in the frequency domain. The use of the butterfly FFT can result in incorrect values for the spectral intensities, because the read out cannot be done at exactly the expected frequency value. All experiments, presented in this work, are performed with more sophisticated algorithms to prevent the problems of the simple butterfly FFT-algorithm. However it is important that the applied FFT algorithm does no automatically fill data points with zeroes [Bracewell 86, Ramirez 85] to reach the required amount of 2^n points before performing the FFT.

2.3.2   Application of Fourier transformation on the stress signal

Materials can be analysed mechanically by a huge amount of analysis methods [Wilhelm 02]. One of those methods is shear rheology, which can examine the mechanical behaviour under shear stress. Elasticity and viscosity are the properties that are measured with rheology. Elasticity is accessible via Hooke's law equation (2.15). The viscosity can very easily be described by Newtons equation (2.1):

$$\frac{F}{A} = \sigma = \eta(\frac{v}{d}) = \eta\dot{\gamma}.$$

Here in an ideal case the force F is acting on two parallel and planar plates with an area A separated by a distance d. The intermediate space is filled with the sample. The stress σ applied is proportional to the shear rate $\dot{\gamma}$. Under steady conditions the polymer has a constant proportionality factor called viscosity η. For a Newtonian fluid it is independent of the applied shear rate $\dot{\gamma}$. In the opposite case, when the viscosity changes as a function of the shear rate $\eta(\dot{\gamma})$, it is called non-Newtonian or non-linear behaviour. We now focus on the simple case, where a single harmonic excitation is applied by a sinusoidal strain. If this sinusoidal excitation is applied with a frequency ω_1 in a symmetric system the mathematical description gets quite simple. The mathematical description is expressed in terms of a differential equation, based on the force balance of kinetic, frictional, potential and the applied force for a controlled stress rheometer [Wilhelm 99]. The mathematical solution of this differential equation is given by a single harmonic function with the excitation frequency ω_1:

$$\gamma(t) = \gamma_0 e^{(i(\omega_1 t + \delta))}. \tag{2.52}$$

In the linear regime see chapter 2.1, the response has an amplitude γ_0 and the phase is shifted by δ. Equation (2.52) can be expressed in terms of its real (G') and imaginary (G") part. In that form physical properties like relaxation times and phase transitions can be extracted due to the dependence of the moduli G' and G" on the frequency, and the temperature of the sample. In the case that the viscosity becomes a function of the applied shear rate $\eta(\dot{\gamma})$, or the elastic modulus a function of the elongation, the solution equation (2.52) is no longer valid. This happens in the non-linear regime. Under the assumption that under

periodic conditions instantaneous adjustment of the viscosity towards the applied
shear rate occurs, the viscosity is due to the symmetry independent of the shear
direction and therefore only a function of the absolute shear rate [Wilhelm 99]:

$$\eta = \eta(\dot{\gamma}) = \eta(-\dot{\gamma}) = \eta(|\dot{\gamma}|). \tag{2.53}$$

The viscosity can be expanded via a Taylor series with respect to the shear rate,
when under oscillatory shear only small non-linear effects are detected. The Taylor expansion for the viscosity at small shear rates is given in equation (2.54)
with the complex coefficients η_0, a and b under oscillatory shear. These complex coefficients can induce phase shifts with respect to the applied frequency
[Wilhelm 99].

$$\eta = \eta_0 + a\,|\dot{\gamma}| + b\,|\dot{\gamma}|^2 + \dots \tag{2.54}$$

If the applied movement is a harmonic oscillation, strain and strain rate are described as the following:

$$\gamma = A_0 sin\omega_1 t, \tag{2.55}$$

$$|\dot{\gamma}| = |\omega_1 A_0 cos\omega_1 t| = \omega_1 A_0\,|cos\omega_1 t|. \tag{2.56}$$

In the next step the absolute value of shear rate signal is treated by Fourier analysis (see chapter 2.3.1) leading to a time dependence of $\dot{\gamma}$ as a sum of different
harmonics [Ramirez 85]:

$$|\dot{\gamma}| = \omega_1 A_0 \left(\frac{2}{\pi} + \frac{4}{\pi} \left(\frac{cos2\omega_1 t}{1\cdot 3} - \frac{cos4\omega_1 t}{3\cdot 5} + \frac{cos6\omega_1 t}{5\cdot 7} - \dots \right) \right) \tag{2.57}$$

$$|\dot{\gamma}| \propto a' + b'cos2\omega_1 t + c'cos4\omega_1 t + \dots \tag{2.58}$$

In equation (2.57) and also in its simplified version equation (2.58) the dependence of the absolute shear rate respectively the viscosity (see equation (2.54))
towards even respectively the odd higher harmonics becomes visible. The insertion of equation (2.57) and equation (2.54) into Newtons equation (2.59) leads
to equation (2.61) [Wilhelm 99]:

$$\sigma \propto \eta\dot{\gamma}, \tag{2.59}$$

$$\sigma \propto \left(\eta_0 + a\,|\dot{\gamma}| + b\,|\dot{\gamma}|^2 \dots \right) cos\omega_1 t, \tag{2.60}$$

$$\sigma \propto [\eta_0 + a(a' + b'cos2\omega_1 + c'cos4\omega_1 \dots)$$
$$+ b(a' + b'cos2\omega_1 + c'cos4\omega_1 \dots)^2 \dots]cos\omega_1 t \tag{2.61}$$

By using addition theorems for trigonometric functions equation (2.61) is simplified and resultantly written as a sum of even harmonics:

$$\sigma \propto [a'' + b'' cos2\omega_1 t + c'' cos4\omega_1 t + ...] cos\omega_1 t. \qquad (2.62)$$

By multiplying the term in the brackets of equation (2.62) with the term $cos\omega t$ the force finally depends exclusively on a sum of odd harmonics:

$$\sigma \propto Acos\omega_1 t + Bcos3\omega_1 t + Ccos5\omega_1 t + ... \qquad (2.63)$$

with A, B and C being complex numbers. In a last step the non-linear torque signal is analysed towards frequency components by Fourier transformation [Bracewell 86, Ramirez 85, Wilhelm 99]. The different frequencies, now presented in a frequency spectra, show peaks at exact the excitation frequency of the fundamental or the higher harmonics of ω_1. Each peak consists of two parts of information, first the intensity of the peak I_n (magnitude), and secondly its corresponding phase Φ_n see chapter 2.3.3.2. The parameter n ($n = 1, 3, 5, 7, 9...$) denotes the odd multiple of the excitation frequency.

2.3.3   Methods to measure and quantify non-linearity

After introducing the basic principles of the FT-rheology two ways to measure
or quantify the non-linearity are described. One is the magnitude of the higher
harmonics, especially the intensity of the third harmonic, and the other is the
phase of the higher harmonics. In the following both quantities are presented.
These considerations are rephrased from the literature due to their general content [Neidhöfer 03a, Wilhelm 02]

2.3.3.1   Magnitude $\frac{I_n}{I_1}$ to quantify non-linearity

One way to quantify the amount of non-linearity is the ratio of intensities of the
n^{th}-harmonic ($I_{n\omega_1}$) to the 1st-harmonic ($I_{1\omega_1}$) of the magnitude spectra after the
FT [Wilhelm 02]:

$$\frac{I_n}{I_1} = \frac{I_{n\omega_1}}{I_{1\omega_1}}, \tag{2.64}$$

here the number $n = 2, 3, 4, 5...$ gives the order of the higher harmonics. In
the non-linear regime strong strain softening behaviour can be found. The step
function represents the case of the maximal reachable non-linearity [Wilhelm 02]
respective maximal strain softening behaviour. Based on a Carreau or an Ostwald-
de-Waele model,

$$\eta = \eta_0 \left[\beta \left| \dot{\gamma} \right| \right]^{-\alpha} = \frac{\eta_0}{[\beta |\dot{\gamma}|]^{\alpha}}, \tag{2.65}$$

and under the assumption that α can reach up to 1 for maximum shear thinning
this results into:

$$\sigma = \eta \dot{\gamma} = \eta_0 \frac{\dot{\gamma}}{\beta |\dot{\gamma}|} = \frac{\eta_0 \gamma_0 \omega_1 cos(\omega_1 t)}{\beta \gamma_0 \omega_1 |cos(\omega_1 t)|} = \frac{\eta_0 cos(\omega_1 t)}{\beta |cos(\omega_1 t)|}. \tag{2.66}$$

By describing the step function with a Fourier series, the maximum values of the
intensities for the different amplitudes are extracted from the following equation:

$$I_{(\omega_1)} \propto \frac{4}{\pi} \left[sin\omega_1 t + \frac{1}{3} sin3\omega_1 t + \frac{1}{5} sin5\omega_1 t \right]. \tag{2.67}$$

The maximum intensity of the harmonics is then given by:

$$\frac{I_{n\omega_1}^{\infty}}{I_{\omega_1}} = I_{\frac{n}{1}}^{\infty} = \frac{1}{n}. \tag{2.68}$$

The maximum value for the intensity of the 3rd harmonic is calculated to be 33
% for the presented assumption. This step function has a periodicity of $T = \frac{2\pi}{\omega_1}$.

Often the assumption that $\alpha = 1$ is not correct. Experimentally α ranges from 0.7 up to 1. Therefore the here predicted maximum value for the intensity of the 3rd harmonic is not always achieved.

To visualise the intensities of the higher harmonics, the ratio of $\frac{I_3}{I_1}$ is plotted as a function of the strain amplitude γ_0 [Wilhelm 98, Wilhelm 00]. To describe this behaviour a function was developed by Wilhelm et al. [Wilhelm 02]. The intensity of the 3rd harmonic has a maximum value, and even at small strain amplitudes a minor non-linear response occurs. An increase of the intensity of the 1st the 3rd harmonic at small shear rates according to γ_0^1 and γ_0^3 [Helfand 82, Pearson 82] results in an increase of the relation $\frac{I_{\omega_3}}{I_{\omega_1}} \propto \gamma_0^2$. At large strain amplitudes a crossover to a plateau behaviour of $\frac{I_3}{I_1}$ is found. This behaviour can be described by the following equation:

$$\frac{I_3}{I_1}(\gamma_0) = A\left[1 - \frac{1}{1+(B\gamma_0)^C}\right], \qquad (2.69)$$

with the parameter A standing for the maximum of the intensity of $\frac{I_3}{I_1}$, and B which is the pivot point of the power law dependency, given by the parameter C.

2.3.3.2   The phase Φ_n to quantify non-linearity

The second parameter giving information about the non-linearity is the phase Φ_n. The phases of the 3rd or of the higher harmonics have a great influence on the symmetry of the resulting time domain signal. Depending on the value of the phase a symmetry loss, visible in the shift of the maxima and minima, is observed. One way to visualise and quantify this symmetry loss is by plotting the shear strain versus the shear stress in so-called Lissajou figures [Giacomin 98]. In the linear regime the shape of the Lissajou figures only depends on the phase shift δ between the strain and stress signal. In the case of $\delta = 0°$ the Lissajou figure shows a diagonal line, whereas in the case of $\delta = 90°$ a perfect circle is achieved. For all other values of δ the Lissajou figures result in an elliptical shape. The difficulty of this method is the influence of higher harmonics on these figures. The higher harmonics change the shape of the ellipses. It is therefore very difficult to quantify the contributions of the phase difference from those of the intensity higher harmonics. As a consequence the phase from the FT spectra is used as a quantification method. To get a comparable rheological information a linear phase correction of the phases of the higher harmonics (Φ_3, Φ_5, Φ_7, Φ_9) with respect to the first phase Φ_1 [Neidhöfer 03a] is done. Therefore equation (2.63) has to be rewritten to include the different phases:

$$\sigma(t) = I_1 cos(\omega_1 t + \varphi_1) + I_3 cos(\omega_3 t + \varphi_3) +$$
$$I_5 cos(\omega_5 t + \varphi_5) + I_7 cos(\omega_7 t + \varphi_7) + ... \tag{2.70}$$

The time shift, that is done in the next step, would be equal to a specific trigger signal on the original mechanical data in the time domain. A substitution of t with $t' - \frac{\varphi_1}{\omega_1}$ is done to receive equation (2.71):

$$\sigma\left(t' - \frac{\varphi_1}{\omega_1}\right) = I_1 cos\left(\omega_1\left(t' - \frac{\varphi_1}{\omega_1}\right) + \varphi_1\right) + I_3 cos\left(\omega_3\left(t' - \frac{\varphi_1}{\omega_1}\right) + \varphi_3\right) + \ldots$$
$$= I_1 \cos(\omega_1 t') + I_3 \cos(3\omega_1 t' + (\varphi_3 - 3\varphi_1)) + \ldots$$
$$\tag{2.71}$$

with the shift factor $-\frac{\varphi_1}{\omega_1}$. Now a relation between the higher harmonics phases to the first phase is extracted, that leads to [Neidhöfer 03a]:

$$\Phi_n = \varphi_n - n \cdot \varphi_1. \tag{2.72}$$

The phase Φ_n can vary from 0 ° to 360 ° ($\Phi_n \in [0°, 360°]$). After introducing the phase analysis as a tool for quantification of the non-linear behaviour a visualisation for this tool is shown to deepen the understanding of this method [Neidhöfer 03b]. In Fig. 2.8 the dependence of the time domain data on the relative phase of the 3rd harmonic is shown. Mathematically two pure cosine functions are added with the frequencies ω_1 and $3\omega_1$. In Fig. 2.8 the amplitudes have a ratio of $\frac{A_1}{A_3} = \frac{1}{0.1}$ and the phase of the 1st harmonic stays constant whereas the phase of the 3rd harmonic φ_3 differs in the range from $\varphi_3 = 0$ over $\varphi_3 = \frac{\pi}{2}$ to $\varphi_3 = \pi$. In Fig. 2.8 plot A the strain softening case is visible. Here the phase of the 3rd harmonic is out-of-phase at an exact phase shift of $\varphi_3 = \pi$ corresponding to 180 °. In case B both the phases, φ_1 and the φ_3, are in-phase, namely 360 ° or 2π. In all other cases the maxima and minima of the resulting time domain signal are either shifted to the left or to the right. In case that the phase is smaller than 180 ° the shift is to the left (see Fig. 2.8 plot C) and to the right if it is bigger than 180 °. The trigonometric function sine or cosine used to shift the higher phases relative to the first of the time domain signal is furthermore responsible for the absolute value of the higher phases. Another solution for this problem was described elsewhere [Giacomin 98]. Here the phase difference of the response φ_n is related to the phase of the excitation δ_γ resulting in:

$$\Delta_n = \varphi_n - n\delta_\gamma. \tag{2.73}$$

In this case the phase difference, quantifying the thinning and thickening behaviour, is not independent of the $tan\delta$, representing relation of the viscous to the elastic contribution of the viscosity.

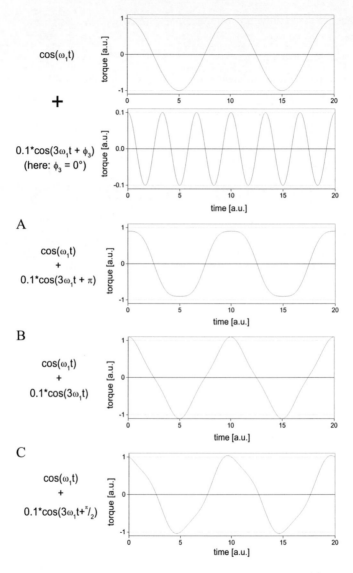

FIG. 2.8: The influence of the phase on the shape of the time domain signal is presented here. Two cosine terms for the fundamental and the third harmonic are added with three different phases at $t = 0$ of the 3rd phase: in-phase (0 °), out-of-phase (180 °) and with a phase of 90 °. In plot A a strain softening, in plot B a strain hardening behaviour is found. In plot C the maxima and minima are shifted to the left [Neidhöfer 03b].

2.4 Dilute dispersions of nonspherical Particles

Non-spherical particles can produce much stronger elastic effects (also larger normal forces) under a flow field, than solutions with spherical particles with similar volume fractions [Larson 99]. The particles, with the simplest non-spherical geometry, have a rotational symmetry axis and are therefore called axisymmetric particles such as rods, disks. An axisymmetric ellipsoid has two principal axes of equal length. Such objects are added in applications to solutions due to their ability of increasing the viscosity or due to the addition of strength to solids such as fiber-reinforced concrete or fiber-reinforced molded plastic parts.

2.4.1 Semi-dilute dispersions of Brownian rods

In this chapter theories for the concentration and for the mechanical behaviour under shear of dispersions of Brownian rods are introduced. A solution of long axisymmetric particles with an aspect ratio of more than 50 are experimentally found as dilute, only if the concentration is below $1\ \%$ by volume [Mori 82]. A requirement for diluteness is the possibility for the rods to rotate freely without being impeded by neighbouring rods. Therefore each rod has to have a free volume of about L^3 with L the length of the rod. Thus rod-rod interactions should be expected when the number concentration of rods c reaches a value equal to L^{-3}. Experiments found the dilute to semi-dilute transition to occur at 30 times this value [Mori 82]. Apparently rods are able to evade other rods that enter their sphere of rotation. Therefore the transition from dilute to semi-dilute behaviour is found at $c^* = \frac{30}{L^3}$. Here c^* is the number concentration at the transition from the dilute to the semi-dilute regime. The volume fraction of the rods φ_{rod} is given by:

$$\varphi_{rod} = \pi d^2 L \frac{c}{4}, \tag{2.74}$$

with d the diameter, L the length of a rod, and c the number concentration of rods. Using equation (2.74) with the relation $c^* = \frac{30}{L^3}$, the cross-over from dilute to semi-dilute occurs in terms of volume fraction at:

$$c^* = \frac{30\pi d^2}{4L^2} \approx 24 \left(\frac{d}{L}\right)^2. \tag{2.75}$$

For an aspect ratio of $\frac{L}{d} = 50$ this cross-over occurs at a volume concentration of $c^* = 1\ vol.\%$. For particles like the FD-virus (see chapter 5) (aspect ratio $\frac{L}{d} = $

100) this occurs already at a concentration of $c^* = 0.24\ vol.\%$. At concentrations
of shortly above c^* the static properties are almost unchanged compared to the
dilute regime. A newly added rod will have a negligible probability of intersecting
with other rods. The cross-over from the semi-dilute to the concentrated isotropic
regime is reached, when rods start to have difficulty in packing isotropically due
to excluded volume interactions. The cross-over from semi-dilute to concentrated
isotropic occurs at the number concentration c^{**} or the volume fraction φ^{**} that
are given by [Doi 86]:

$$c^{**} \approx \frac{1}{dL^2} \quad \text{or} \quad \varphi^{**} \approx \frac{\pi d}{4L}. \tag{2.76}$$

At higher concentrations the excluded volume interactions lead to the formation
of a nematic liquid crystalline state. Due to the excluded volume theory of On-
sager, the highest concentration, where there is still an isotropic phase, is at φ_1
$\approx 3.3\ \frac{d}{L}$. For an aspect ratio of $\frac{L}{d} = 100$ this results in a volume fraction of $\varphi_1 =$
3.1 %. In cases where the rod length is longer than the persistence length equa-
tion (2.76) changes to $\varphi_1 \propto \frac{d}{\lambda_p}$. At higher concentrations, transitions to positional
ordered liquid crystalline phases (e.g. smectic) are observed. The rotational dif-
fusion coefficient for these rods can be derived from Doi-Edwards theory. For rods
the basic idea of the theory is based on a cage acting like a cylinder corresponding
to the tube in the polymer model. The radius of the cage is given by $a \sim \frac{1}{\nu L^2}$. The
difference to 'the tube in the polymer' model is, that the rod must remain completely oriented as long as a part is still inside the cage, whereas a polymer chain
can change its orientation as soon as it is only partly outside the tube. The rod
can change its orientation only by a small angle ϵ (in rad) $\approx \frac{a}{L} \approx \frac{1}{\nu L^3}$ [Doi 86].
The time the rod needs to get totally disoriented is given by:

$$\frac{\tau_D}{\epsilon^2} \approx \frac{\tau_D L^2}{a^2} \approx \tau_D \left(\nu L^3\right)^2. \tag{2.77}$$

The rotational diffusion coefficient D_r, being the inverse of the relaxation time, is
given by [Doi 86]:

$$D_r = \beta D_{r0}(\nu L^3)^{-2}, \tag{2.78}$$

with D_{r0} as the dilution state rotational diffusion coefficient. The dimensionless constant β is quite large (around $1,350$) for perfectly rigid rods [Teraoka 89,
Bitsanis 90]. Under flow the orientation distribution of the rods becomes

anisotropic and the radius of the cage increases to $a \propto \frac{1}{sin\theta}$ with θ the orientation angle of a caged rod with respect to the trapped test rod.

A theory for the description of the mechanical behaviour under shear is introduced here. The results from this theory will later be compared with experimental results. The basic ideas of a model newly developed by Dhont et al. [Dhont 03] for long rods is presented here. This model is an extension of the theory from Doi, Edwards and Kuzuu [Doi 78, Doi 86, Doi 81, Kuzuu 83]. This theory includes two very important quantities for very long and thin rods: The concentration of the rods, and the orientational order parameter tensor $S = \langle \hat{u}\hat{u} \rangle$, measuring the orientational order. Here $\hat{u}$ is the unit vector along the long axis of the uniaxial rod. An equation of motion for the tensor $\underline{S}$ will be derived and solved. Using this result the stress tensor is calculated. In order to obtain an analytical result for the leading term of the shear rate dependence of the zero shear viscosity and normal stress differences, the orientational order parameter tensor is expanded up to the third power in the shear rate. For details concerning the calculations the interested reader is referred to the literature [Dhont 03]. The extra stress tensor is then given by:

$$\underline{\underline{\sigma}} = 2\eta^{eff}\dot{\gamma}\underline{\underline{E}} + \eta_0 \frac{1}{120} \times \frac{\dot{\gamma}^2}{D_r^{eff}}\alpha\varphi \begin{pmatrix} 19 & 0 & 0 \\ 0 & -11 & 0 \\ 0 & 0 & -8 \end{pmatrix}, \tag{2.79}$$

α is defined as:

$$\alpha = \frac{8}{45} \frac{(\frac{L}{d})^2}{ln(\frac{L}{d})}. \tag{2.80}$$

Note, that D_r^{eff} is a collective (effective) diffusion coefficient, which describes the decay (or initial growth) of a very small perturbation of an initially isotropic state:

$$D_r^{eff} = D_r \left[1 - \frac{1}{5}\frac{L}{d}\varphi\right], \tag{2.81}$$

and $\hat{E} = \frac{1}{2}[\hat{\Gamma} + \hat{\Gamma}^T]$ the symmetric part of the velocity gradient tensor. The ratio of $\dot{\gamma}$ and D_r^{eff} is called the dressed rotational Péclet number. For low shear rates the suspension shear viscosity η^{eff} is given by [Dhont 03]:

$$\eta^{eff} = \eta_0 \left[1 + \left(1 - \frac{1}{50}\left(\frac{\dot{\gamma}}{D_r^{eff}}\right)^2\right)\alpha\varphi + \frac{1}{1500} \times \frac{\dot{\gamma}^2 D_r}{(D_r^{eff})^3}\alpha\frac{L}{d}\varphi^2\right]. \tag{2.82}$$

In the case of higher shear rates it is given by:

$$\eta^{eff} = \eta_0 \left[1 + \left(1 - \frac{1}{50} \left(\frac{\dot{\gamma}}{D_r} \right)^2 \right) \alpha\varphi - \frac{11}{1500} \times \left(\frac{\dot{\gamma}}{D_r} \right)^2 \alpha \frac{L}{d} \varphi^2 \right], \qquad (2.83)$$

with the concentration:

$$\frac{L}{d}\varphi. \qquad (2.84)$$

For shear rate $\dot{\gamma} = 0$ equation (2.83) is simplified to see [Dhont 03]:

$$\eta^{eff} = \eta_0(1 + \alpha\varphi). \qquad (2.85)$$

This equation is the analogue of Einstein's equation for the viscosity of dilute dispersions of spheres:

$$\eta^{eff} = \eta_0(1 + \frac{5}{2}\varphi). \qquad (2.86)$$

Furthermore the viscoelastic behaviour, visualised by the viscosity at higher strain amplitudes, is of interest. To be able to plot the in-phase and the out-of-phase part of the viscosity as a function of the dimensionless Deborah number Ω equation (2.87) is derived:

$$\underline{\sigma} = 2\dot{\gamma}_0\hat{E}[\eta'\cos\omega t + \eta''\sin\omega t], \qquad (2.87)$$

with the dimensionless Deborah number:

$$\Omega = \frac{\omega}{D_r}. \qquad (2.88)$$

For more details the reader is referred to the literature [Dhont 03].

The dissipative η' and the storage shear viscosity η'' are given by:

$$\eta' = \eta_0 \left[1 + \left(\frac{1}{4} + \frac{9}{2} \frac{F\frac{\Omega^{eff}}{6}}{\Omega^{eff}} \right) \alpha\varphi \right], \qquad (2.89)$$

$$\eta'' = \eta_0 \frac{3}{4} F \left(\frac{\Omega^{eff}}{6} \right) \alpha\varphi, \qquad (2.90)$$

with the substitution function F:

$$F(\Omega^{eff}) = \frac{\Omega^{eff}}{1 + (\Omega^{eff})^2}, \qquad (2.91)$$

and Ω^{eff} the dimensionless concentration dependent rotational Deborah number:

$$\Omega^{eff} = \frac{\omega}{D_r^{eff}}. \qquad (2.92)$$

For higher shear rates, when the behaviour reaches the non-linear regime, the stress can be calculated via equation (2.87):

$$\sigma = 2\dot{\gamma}_0 \hat{E} \sum_{n=0}^{\infty} [\eta'_n cosn\omega t + \eta''_n sinn\omega t]. \tag{2.93}$$

The bare rotational Péclet number is defined by:

$$Pe_r = \frac{\dot{\gamma}}{D_r} \tag{2.94}$$

Chapter 3

Experimental Issues

In this section experimental issues like the FT-rheological set-up, some information about Couette cells, and the zero shear viscosity η_0 are addressed.

3.1   Experimental set-up of the FT-rheology

The FT-rheology set up consists of a Rheometrics Scientific Advanced Rheometer Expansion System (ARES) and two computers. This rheometer is a strain controlled rheometer. In this context an information about rheological techniques should be given. These two different shear methods, steady and dynamic, are performed on a rotational rheometer within this work. There are two different possibilities of how these rheometers are controlled. In our case a deformation is applied on the sample and a torque is measured. Given the deformation and therefore the shear rate, the rheometer is called a CR-rheometer, controlled rate. In the other case a defined shear stress is applied to the sample. The measured property here is either the deformation or the shear rate. This method is called CS-controlled rheometer, which means controlled stress [Gedde 95, Schramm 95]. The rheometer is equipped with a dual range Force Rebalance Transducer (100 $FRTN1$) and is capable of measuring torques ranging from 0.004 mNm to 10 mNm. It is furthermore equipped with a high resolution (HR) motor. This motor can apply frequencies ranging from 10^{-5} to 500 $\frac{rad}{s}$ and a deformation amplitude ranging from 0.005 to 500 $mrad$. The oscillation frequency can be varied from 0.001 up to 100 $\frac{rad}{s}$. It is furthermore equipped with a water bath, which has a temperature range from $-20\ ^\circ C$ up to 95 $^\circ C$. On the backside of the machine

41

it is connected with a serial cable to a PC, which controls the rheometer. Additionally three BNC cables are connected to a second computer where the data for the FT-rheology is acquired and analysed (see Fig. 3.1). For the further data treatment the raw signals are read out and the analog rheological signals are then digitised with a 16-bit ADC (the dynamic range is $1 : 65,536$). A PCI-MIO-16XE-10 card from National Instruments, USA is used, which has a maximum sampling rate of $100\ kHz$ and is capable of multiplexing up to 16 channels. It can simultaneously acquire and transfer the data to the PC memory by data-buffering techniques. Therefore the rheological data is intrinsically synchronised. The 40 μs interchannel delay (time between consecutive data points) between the four channels is relatively insignificant compared to the timescale of rheological experiments. The three channels are namely strain, torque and normal force. In the experiments performed in this thesis an increase of the $\frac{S}{N}$ ratio by a factor of 3 to 5 was achieved. These values are smaller than the theoretical $\frac{S}{N} \propto \sqrt{n}$, due to practical reasons based on the set-up. The ADC-card acquires the time data at the highest possible sampling rate, and the data is then on-the-fly preaveraged to reduce random noise. The method is described in more detail in [Dusschoten 01]. In the experiment the data is acquired via a k-bit analog-to-digital converter card (ADC-Card). The ADC-card has a sampling rate of $100,000$ $\frac{1}{s}$. This means that the analog time data $s(t)$ is measured discretely in $100,000$ time steps and is then converted to digital data. In our case a 16-bit ADC-card is used. The "16-bit" give an information of the ability of the card to discriminate the intensity of the signal. It can discriminate $2^{16} - 1$ steps in intensity. Furthermore the 16-bit is, in connection with the measurable voltage range, a measure for the minimum detectable intensity of weak signals. The bigger the bit number, the smaller the minimal detectable intensity. If the bit number is to small, the ADC-card is the limiting factor towards a small $\frac{S}{N}$ ratio. For more details the reader is referred to [Claridge 99, Homans 89, Skoog 92]. After acquisition the time data is averaged using of a home-written LabVIEW software program [Dusschoten 01, Neidhöfer 03a]. The FT-Analysis is then performed with a different home-written LabVIEW program the analyser software see Appendix B.1. For the stress and strain signals a simple averaging is conducted on-the-fly in the time domain [Dusschoten 01]. The rheological data is then averaged. The smallest time steps are $50\ ms$ equaling $1/50,000$ $\frac{1}{s}$. For information about any used

LabVIEW software programs see Appendix B.

For watery solutions three different types of Couette-cells are used: a) Couette cell with a bob diameter of 32 mm and a cup diameter of 34 mm, b) Couette cell with a bob diameter of 16 mm and a cup diameter of 18 mm, and c) double wall Couette Cell. The cup has a slit of the radius between 27 mm and 34 mm. The corresponding bob has a thickness of 3 mm in the radii between 29 mm and 32 mm. The normal force is not used in my experiments due to the fact that in Couette geometries the normal force measurement does not make sense. In Couette Cells the sample will just push up to the samples surface which it restricted by the surrounding but not by the geometry.

For the measurements an appropriate amount of the sample is filled into the selected Couette cell. A major problem is the evaporation of the water. A solvent trap is used for this purpose. This trap is equipped with a sponge drawn with water. Additionally dodecane $C_{12}H_{26}$ covers the solution to prevent an exchange with the atmosphere above. Dodecane is non-polar and has, with a viscosity of 1.383 $mPas$ [Lide 96], a lower viscosity than the samples with a high particle loading. Typically 40 cycles have been acquired, while a scanning rate of 50,000 $\frac{1}{s}$ for each channel and an oversampling of 1,000 are typically selected. This results

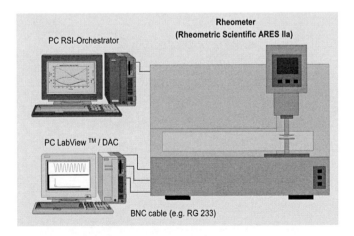

FIG. 3.1: The experimental set-up of the FT-rheology. The normal ARES-Rheometer plus the addition with a computer including the LabVIEW software.

in 50 data points per second (respectively for a sine wave with a frequency of $1Hz$), which is enough for the purposes of my measurement.

3.2 Couette Cell

The use of the correct geometry of the rheometer is of great interest for the measurement within this Ph.D. thesis. Here, several geometries should be mentioned such as plate-plate, cone-plate, and concentric cylinders. Except for the plate-plate geometry the sample has a spatially homogeneous shear rate. For samples with a lower viscosity the concentric cylinders are used, because they have a bigger contact surface which is useful for liquid samples. They are the choice for the samples investigated within this dissertation. Concentric cylinders were first used by Maurice Couette [Couette 90] in the nineteenth century. These concentric cylinders always consist of an outer cup connected to the bottom and an inner bob connected to the top. At this point two different experimental set-ups are introduced (see Fig. 3.2). The torque signal is measured in both cases at the top, whereas the rotation is done differently. In the first set-up the inner bob rotates and the cup is fixed. This set-up, named after Searle, has the disadvantage to create larger secondary flows like the Taylor vortices Fig. 3.2. These are small axisymmetric cellular motions that dissipate energy and therefore increase the measured torque. A criterion for the occurrence of Taylor vortices [Taylor 23] is the Taylor number defined as:

$$Ta = \frac{\rho^2 \omega_1^2 (R_0 - R_i)^3 R_i}{\eta(\dot{\gamma})^2},$$

(3.1)

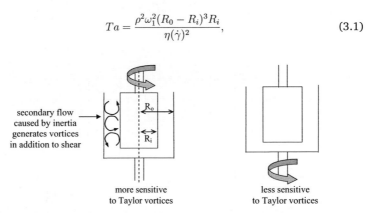

secondary flow caused by inertia generates vortices in addition to shear

more sensitive to Taylor vortices

less sensitive to Taylor vortices

FIG. 3.2: A visualisation of Taylor vortices in parallel cylinders. Also shown is the Couette design, which is less sensitive to evoke Taylor vortices.

with the outer radius R_0, and the inner radius R_i. When the Taylor number exceeds $3,400$ instabilities occur. In the second set-up the outer bob rotates and Taylor vortices are highly suppressed. In this case the flow is stable until turbulences occur at high Reynold's numbers N_{Re} exceeding $50,000$ [vanWazer 63], where the Reynold's number is given for a Couette set-up as:

$$N_{Re} = \frac{\rho \omega_1 R_0 (R_0 - R_i)}{\eta_s}. \tag{3.2}$$

In a Couette cell the rotation frequency ω_1 is zero at the inner radius R_i whereas at the outer radius R_o it has its maximal value for ω_1. The shear stress is given by:

$$\sigma = f(r) = \frac{M_t}{2\pi R_0^2 h}. \tag{3.3}$$

Here h is the height of the cell and M_t the measured torque, which can be calculated under Newtonian flow as:

$$M_t = \frac{4\pi R_i^2 \cdot R_0^2 \cdot h}{R_0^2 - R_i^2} \cdot \eta \cdot \omega_1 = K_c \eta \omega, \tag{3.4}$$

where K_c is a constant of the geometry. By inserting equation (3.4) in equation (3.3) the shear stress is given by:

$$\sigma_{12} = \frac{M_t}{2\pi \cdot R_i \cdot R_0 \cdot h}. \tag{3.5}$$

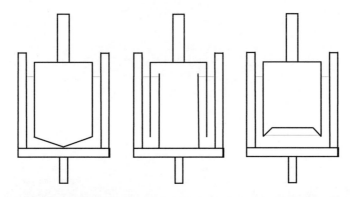

FIG. 3.3: Three different designs of Couette cells: Mooney-Ewart on the left, Double Wall Couette in the middle and the Haake design Couette on the right with the low friction air reservoir at the bottom.

The shear rate and the viscosity are as follows:

$$\dot{\gamma} = \frac{2R_i \cdot R_0}{R_0^2 - R_i^2} \cdot \omega_1, \tag{3.6}$$

$$\eta = \frac{(R_0^2 - R_i^2)}{4\pi \cdot hR_0^2 \cdot R_i^2} \cdot \frac{M_t}{\omega_1}. \tag{3.7}$$

The influence, which the bottom of the bob and the bottom of the cup have on the torque, must be corrected, since at the bottom of the cylinder there is also shear flow. Three different basic designs of Couette-bobs have been developed to counter the end effects (see Fig. 3.3). One features a conical bottom. With proper choice of the cone and angle, the shear rate at the bottom agree with that in the narrow gap on the sides. The next design has a thin rotating inverted cup and is called double wall Couette (see Fig. 3.4). In general it allows a better temperature control, due to the large contact surface and a small sample volume. A specific feature of the double wall Couette is that, except for small gaps, the shear rates on the inside and outside of the rotating cup are not equal. The greatest advantage

FIG. 3.4: This picture shows a double wall Couette.

of the double wall Couette, is the small amount of sample needed for its use and the use of twice the contact area to increase sensitivity, for low viscous systems. The third design, the so-called Haake-design, has a recessed bottom, trapping air, which essentially transfers no torque to the fluid. Haake design and double wall Couette have both been used within this Ph.D.

3.3 Zero-shear viscosity η_0

The viscosity as a function of the solid content for a the limiting case of zero shear is called zero-shear viscosity η_0. The simplest dispersions are composed of so-called hard spheres in a solvent of viscosity η_s in which the only interaction between particles are rigid repulsions that occur when particles come into contact. These simple dispersions can already show a complex rheological behaviour. The zero-shear viscosity, the viscosity of a solution extrapolated to no shear, of hard spheres solutions can be calculated at very low volume fractions φ to:

$$\eta_0 = \eta_s(1 + 2.5\varphi). \tag{3.8}$$

This equation is a result of calculations from the viscous dissipation produced by the flow around a single sphere. This calculation was first conducted by Einstein [Einstein 06, Einstein 11] and is experimentally only valid for low volume fractions $\varphi \leq 0.03$, so that the flow field around the sphere is not influenced by the presence of neighbouring spheres. If, at higher particle loadings, two spheres are close enough that the shear field on one sphere is influenced by a second sphere. This gives rise to a proportionality for the viscosity η of φ^2. In the case that three particles interact, this relation changes to φ^3. The effect for two particle interactions was first calculated by Batchelor [Batchelor 71]. When combining the results from Einstein and from Batchelor [Larson 99] a new relation is given that is valid for a volume fraction $\varphi \leq 0.1$:

$$\eta_r = \frac{\eta}{\eta_s} = 1 + 2.5\varphi + 6.2\varphi^2. \tag{3.9}$$

The expansion of equation (3.9) was extended to higher order in φ by a simple effective medium argument of Arrhenius [Arrhenius 17]. The increase in viscosity η can be calculated by adding particles $d\varphi$ to a suspension, treated like a homogeneous medium with the viscosity $\eta(\varphi)$ and can therefore be extended to higher orders in φ. The viscosity in Einstein's equation (3.8) is incrementally increased to:

$$d\eta = 2.5\eta(\varphi)d\varphi. \tag{3.10}$$

After integration the following equation is reached:

$$\eta = \eta_s e^{(\frac{5\varphi}{2})}. \tag{3.11}$$

A similar argument can be made for arbitrarily shaped particles leading to equation (3.12) where $[\eta]$ is the dimensionless intrinsic viscosity, which is the dilute limit of the viscosity increment per unit particle volume fraction, divided by the solvent viscosity:

$$\eta = \eta_s e^{([\eta]\varphi)}, \tag{3.12}$$

$$\eta_{intrinsic} = \frac{\eta - \eta_s}{\varphi \eta_s}, \tag{3.13}$$

$$[\eta] = \lim_{\varphi \to 0} \frac{\eta - \eta_s}{\varphi \eta_s}. \tag{3.14}$$

Equation (3.14) is known as the Staudinger index. Equation (3.12) leads to equation (3.11) if for spheres $[\eta]$ has the value of 2.5 [Amelar 91]. At high sample loading this relation fails due to 'crowding'. 'Crowding' means that there is a correlation for a particle being at a specific position because of the influence of another particle. This problem is circumvented by replacing [Ball 80] the viscosity increment by:

$$d\eta = \frac{[\eta]\eta_s d\varphi}{1 - \frac{\varphi}{\varphi_m}}. \tag{3.15}$$

The viscosity diverges when φ approaches the maximum-packing volume fraction φ_m. Note, that φ_m for ordered dense packing of monodisperse spheres has the value of 0.74, whereas here the value for the disordered case is used. For hard spheres this is known from crystallography [Bernal 60], to be $\varphi_m \approx 0.63$ to 0.64. The integration of equation (3.15) leads to equation (3.16) that is known as the Krieger-Dougherty equation:

$$\eta = \eta_s \left(1 - \frac{\varphi}{\varphi_m}\right)^{-[\eta]\varphi_m}. \tag{3.16}$$

Empirically the theoretical results could be confirmed for dispersions of particles or other shape in 1959 [Krieger 59, Krieger 63]. For non-spherical particle solutions the value of $[\eta]$ increases whereas the value of φ_m decreases when the particle aspect ratio increases [Meeker 97]. In particle solutions, with sterically stabilised particles at volume fractions far from close dense packing (≈ 71 $vol.\%$), the thickness of the sterically layer Δ has an additional influence on the volume fraction [Mewis 89] resulting to:

$$\varphi = \varphi_0 \left(1 + \frac{\Delta}{a}\right)^3, \tag{3.17}$$

with a the radius of the particle and φ_0 the volume fraction of the uncoated particles. Another equation describing the zero-shear viscosity, as a function of the volume fraction is from Quemada [Quemada 78]:

$$\eta_0 = \eta_s \left[1 - \frac{\varphi}{\varphi_{max}} \right]^{-2}, \tag{3.18}$$

with $\varphi_{max} = 0.63$ meaning the volume fraction for the disordered packing of spheres [Rueb 98]. For the high shear viscosity a different value for $\varphi_m = 0.72$ was found [deKruif 85]. The theoretical predictions of Einstein, Krieger-Dougherty and Quemada are plotted in Fig. 3.5. Interesting effects of the viscosity can be observed at very a high loading of bimodal distributions of particle size [Dames 01]. Mainly, the viscosity is higher for mono-modal dispersions. A strong drop in viscosity (more than a decade) can be observed in bimodal solutions with a size ratio $5 : 1$ at a volume fraction above 0.6. At this volume fraction the small particles go into the interstice of the lager ones. This effect is referred to the 'Farris ef-

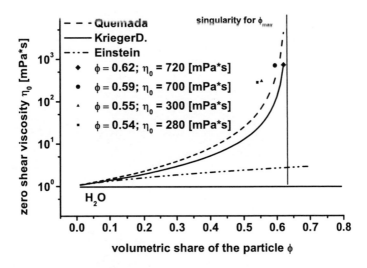

FIG. 3.5: Zero-shear viscosity as a function of volume fraction as described by different models: Einstein and Krieger-Dougherty, and Quemada for a water suspension.

fect' [Larson 99]. This observation has industrial relevance, because dispersions can be produced that have higher sample loading, but less water, and can still be treated like a low solid content dispersion. Stokesian dynamics computer simulations of hard-sphere dispersions give insight into their shear-thinning behaviour [Bossis 89, Phung 96]. At high shear stresses, the contribution of the Brownian motion disappears and leaves only the hydrodynamic contribution. Other simulations [Visscher 94, Phung 96] for volume fractions of $\varphi = 0.30$ and $\varphi = 0.45$ show formations of lines of particles, co-called strings, parallel to the shear flow. Their appearance correlates with the onset of shear thinning and vanishes with the start of shear thickening. These strings, forming superstructures perpendicular to the flow direction in a hexagonal pattern [Laun 92] can be verified by light scattering [Ackerson 88]. In liquids consisting of small molecules, which show shear thinning at high shear rates, these structures can also be found. An even higher drop in viscosity is found in dispersions showing a three-dimensional order like close-packing in face centered cubic (fcc) and / or hexagonal close packing (hcp) with a volume fraction of $\varphi \approx 0.5$. Under shear stress these samples lose their 3D structure and form layers that make a flow easier possible. But before losing their order a yield stress has to be applied. Furthermore similar effects have been found in 'soft' particles or particles that have a grafted surface layer [Mewis 89].

Chapter 4

Dispersions and Synthesis

In this chapter dispersions, their definition, their importance and their synthesis within this dissertation is presented. The definition of a dispersion is a solid dispersed in a fluid, where the solid is not dissolved [Macosko 94]. An emulsion is, in contrast to the dispersion, a liquid dispersed in a liquid. The emulsion polymerisation is based on emulsions. The synthesis leads then to a dispersion. An example for an dispersion is lime milk, where calcium hydroxide is dispersed in water. In contrast to normal milk, which contains fat and protein droplets, where fluid is dispersed in water and therefore called an emulsion. In the fields of polymers the definitions are more complicated. A polymer dispersion could contain polymers that have a more glassy character or viscous fluid character. Furthermore, if a watery dispersion is synthesised via emulsion polymerisation, it is called latex. A typical diameter of the particles for polymer dispersions are in the range of $50\ nm$ $-\ 500\ nm$. Latices with particles size in the range of $50\ nm - 500\ nm$ are called colloids [Xia 00, Li 00]. The systems used in this dissertation were synthesised via emulsion polymerisation. This method was first patented in 1909 in Germany [Hofmann 09]. The breakthrough for this method came many years later during the second world war, when the import of natural rubber to Germany and to the USA was stopped. As an alternative styrene-butadiene-latices were then produced in Germany and the USA. Today latices are widely used in industry for a multitude of applications like textiles, leather, wall paints, paper coatings, as a binding agent in construction industry and also in recently developed research fields like immuno assays [Distler 99]. The emulsion polymerisation is based on a radical polymerisation. It has several advantages compared to bulk or solvent polymerisation. One advantage is that the reaction enthalpy is directly transferred into the reaction medium water and no uncontrolled polymerisation takes place. This heat transfer is highly efficient due to the big interface between the particles and the surrounding water. Due to the high heat capacity of the water no uncontrolled polymerisation can take place. A typical reactor is shown in Fig. 4.1. Normally the reaction temperature lies between 70 °C and 85 °C. The ability of stirring the dispersion, even at high solid content is due to the low viscosity. The cause for it is that the viscosity of the dispersion does not depend on the molecular weight of the polymer but on the medium and the particle interactions. A 30 $wt.\%$ watery dispersion has a viscosity of $1 - 2\ mPas$ which is in the range of the viscosity of water, whereas a 8 $wt.\%$ polystyrene in dioctyl phthalate ($M_n = 200,000\ \frac{g}{mol}$) solution is already a highly viscous medium $\eta(\dot{\gamma} = 0.01) = 700\ Pas$.

In recent years production techniques that allow a synthesis without organic solvents are promoted due to environmental reasons. Water based latex wall paint is dominantly used for indoor applications. In chapter 4.1 synthetic aspects of the

FIG. 4.1: Typical reactor used for synthesising a dispersion.

emulsion polymerisation will be described in a more detail, and characterisation techniques are covered in chapter 4.2. For further information see e.g. [Piirma 82, Gilbert 95, El-Aasser 97].

4.1 Emulsion polymerisation

In the typical emulsion polymerisation the reactive solution is a mixture of surfactant, initiator monomer dissolved in water. A more newly developed method is the 'surfactant free' emulsion polymerisation, where the surfactant is replaced by an ionic monomer like sodium styrene sulfonate. A latex is synthesised according to each different purpose so different additives like rheological modifiers or fungicides are added. The monomers vinyl acetate, acrylates, styrene and butadiene are the basis of the most commonly produced dispersions in industry. In contrast to the main monomer, which is responsible for glass transition temperature, swelling ability and elasticity, the so-called helping co-monomers often play the role of stabilising the dispersion. Mostly their amount does not exceed 5 $wt.\%$. Typical examples for helping monomers are acrylic- or methacrylic acid, creating surface charges on the particles, or bifunctional acrylates and divinylbenzene used as cross-linkers.

Commonly used surfactants are sodium dodecyl sulfate Fig. 4.2 or sodium styrene sulfonate Fig. 4.3. These surfactants create negatively charged particles due to their anionic end groups. For positively charged particles cationic amphiphiles are used, whose hydrophilic part is often quaternary ammonia. A third possibility are non-ionic surfactants, which contain polyethylene oxide units. The polymerisation is started by initiators (e.g. ammonium peroxodiusulfate (APS), potassium peroxodiusulfate (KPS), and azo-*bis*-isobutyrylnitrile (AIBN)) forming radicals while heated up in the range of between 60 $°C$ and 100 $°C$. The cre-

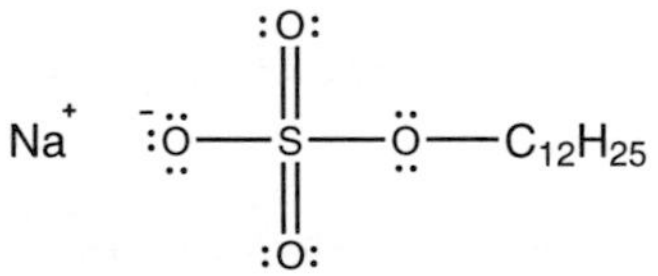

$$Na^+ \quad :\ddot{O}-\underset{\underset{:O:}{\overset{:O:}{\overset{\|}{\underset{\|}{S}}}}-\ddot{O}-C_{12}H_{25}$$

FIG. 4.2: The surfactant sodium dodecyl sulfate, that is frequently used in this work.

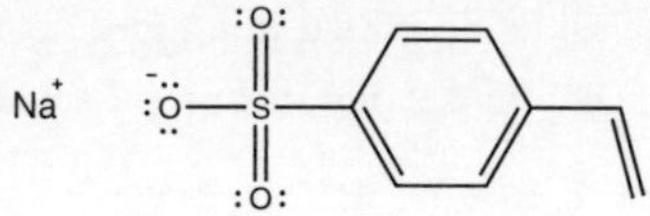

FIG. 4.3: The surfactant sodium styrene sulfonate is used for surfactant-free emulsion polymerisation.

ation of radicals will be described in a more detailed way in chapter 4.1.1. Most significant for the emulsion polymerisation is the creation of micells see Fig. 4.4. These micells are small droplets of monomer (immiscible with water), stabilised by surfactants. The radicals diffuse from the place of their formation in the solution into the micells where the polymerisation reaction takes place. In the course of the reaction the micelle contains more and more polymer and finally it is a polymer sphere dispersed in water. There are different possibilities to perform such a synthesis. One possibility is the batch process, where all ingredients are mixed together and secondly a semi-continuous method where a nucleation phase is followed by a growth phase. In the batch process the whole reactive medium is mixed and then the reaction is accomplished under permanent stirring and heating. It is frequently used to synthesise homo polymeric particles. A drawback is the poor possibility to control the reaction of this exothermal reaction. If the reaction heat is not dissipated, inhomogeneity of the temperature can be found. This becomes a problem when synthesising dispersions with a high solid content, due to the reduced amount of water. Therefore, it is often only used as a first

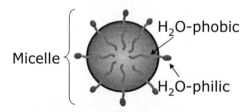

FIG. 4.4: The micelle, a monomer droplet emulsified in water, is stabilised by a surfactant layer. The surfactant arranges itself in such a way that the hydrophobic part points towards the monomer droplet, and the hydrophilic part points towards the water.

step of the semi-continuous method 4.1.2. In the first step of the semi-continuous method seed particles are created. In the second step, the continuous step, the monomer and the other reactive agents are slowly added. So the reactive media is under 'starved-feed' conditions. As a result a better control of the reaction speed, the temperature, the particles size and the particle size distribution is given.

The mechanism of the emulsion polymerisation itself consists of three steps: the nucleation phase, the particle growth and the end of the reaction. The nucleation phase is defined by the formation of seed particles. Two different nucleation mechanisms, which depend on the hydrophobicity of the monomer, are discussed. If the surfactant concentration is above the critical micellar concentration (cmc) or the hydrophobic monomer is used it is called hetero-geneous nucleation [Vanderhoff 85, Song 89]. In a mechanism proposed by Harkins [Harkins 47, Harkins 50] and Smith-Ewart [Smith 48] spherical micells and monomer droplets co-exist. The micellar size depends on the amount of sur-factant and can range between $2\ nm$ up to $30\ nm$. Inside they contain monomer whereas outside there is a layer of surfactant. Some monomers can even be dis-solved in water if it is not too hydrophobic. The dissolved monomer reacts with the added radical initiator to oligomer radicals. The longer the chains grow, the easier they move into micells where the polymerisation is continued. A perma-nent flow of monomer from the monomer droplets guaranties a further poly-merisation. But also radicals can migrate from micells and start new chains [Casey 94, Morrison 94] outside. The chains, growing into spheres, also need surfactant molecules to stabilise their surface. When no free surfactant is left, the nucleation phase is finished. In case the amount of surfactant falls below the cmc or a hydrophilic monomer is used, the nucleation is called homogeneous [Hansen 78]. Oligomer radicals form a primary nucleus after reaching a critical molecular weight. The further growth of the particles, by monomer or oligomer, is then called precursor particle growth. Finally, several precursor particles com-bine to the final particles. The end of the nucleation phase is achieved when there are no surface active oligomers left [Janssen 93]. In the second phase, the par-ticle growth, monomer molecules diffuse from monomer droplets in the solution through the water phase into particles. In the end of this phase no monomer droplets are left in the water, but the only non-reacted monomer is inside the par-ticles. The particles have now reached their final size. The polymerisation speed

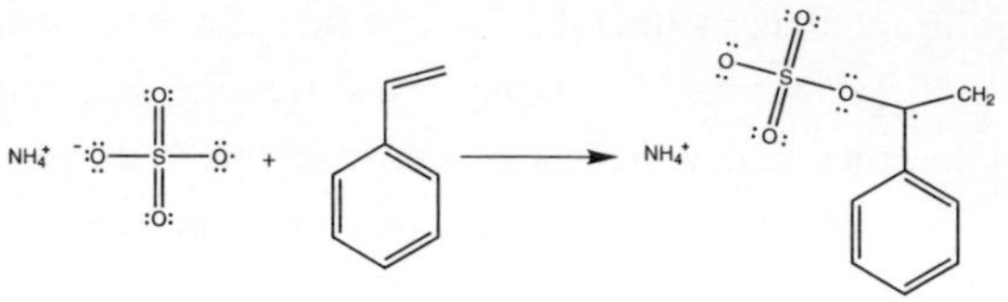

FIG. 4.5: The startreaction of the radical polymerisation is shown here.

is approximately constant during this phase. Finally, at the end of the reaction the surviving monomer inside the particles is polymerised.

4.1.1 Mechanism of the radical polymerisation

A radical polymerisation is a chain reaction which consists of 3 characteristic steps (see above). First the start reaction (formation of the radicals), second the chain growth reaction and finally the termination reaction [Lechner 96]. The start is the formation of a radical. This radical is generated by temperature increase or by UV-light (see Fig. 4.5).

Normally two radicals are created by a symmetrical break of a bond, or the disappearance of nitrogen. Typical radical starters are peroxides (see Fig. 4.6), azo compounds (see Fig. 4.7), hydroperoxides, or organo metal compounds. The radicals are normally created in-situ, meaning the radical concentration is equally spread in the reactive medium. Otherwise the reaction would not be started, due to the recombination of the radicals with the non-reactive surroundings. With the creation of the radical the next step, the startreaction is initiated (see Fig. 4.5). Here the radical covalently binds to the monomer molecule. A new bond between the radical and the monomer is made. The radical electron is now at the end of the newly created molecule. This second step, where the newly formed radical attacks further monomers, is called chain growth reaction (see Fig. 4.8).

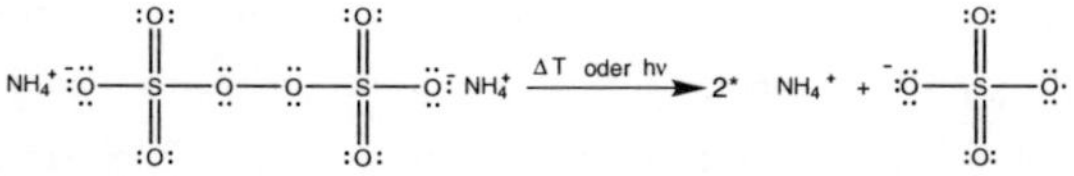

FIG. 4.6: The initiator molecule ammonium peroxodiusulfate, $(NH_4)_2S_2O_8$.

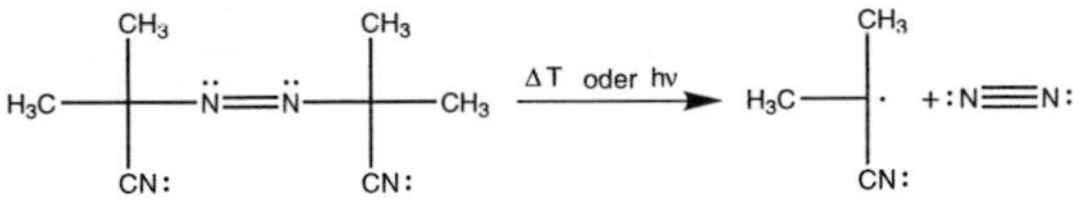

FIG. 4.7: The initiator molecule azo-*bis*-isobutyrylnitrile, AIBN.

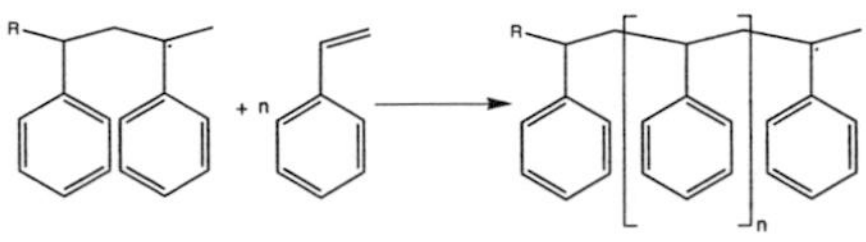

FIG. 4.8: The chain growth reaction is named the polymerisation reaction.

This chain growth is the radical polymerisation process. Finally the reaction stops. Several possibilities can induce the end of the reaction. One possibility is the use of inhibitors or retarders. With the addition of these molecules the reaction is stopped. The reactions conducted within this thesis are stopped via recombination and disproportionation. In the case of the recombination two radicals react, and build one chain of double length (see Fig. 4.9). For reactions with a need for a narrow molecular weight distribution, this kind of stop reaction is undesired. This aspect is generally of less importance in case of the emulsion polymerisation because the created spherical particles have the same diameter. As the sphere diameter does not depend on the chain length a double chain length of two combined radical chains does frequently not matter for the size of the particles. A second termination reaction is the disproportionation. Here two radicals react with each other, in such a way that an electron and a hydrogen are exchanged. The result (see Fig. 4.10) is a creation of a double bond in one molecule

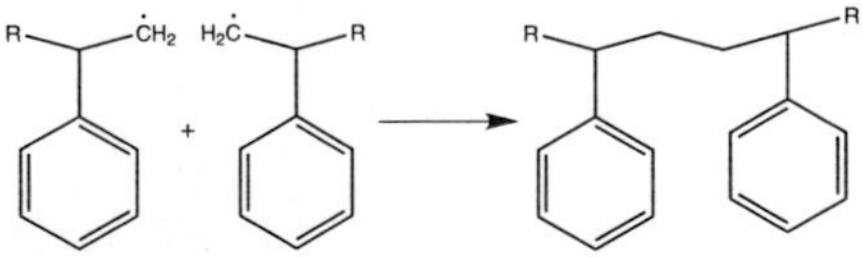

FIG. 4.9: The combination of two radicals terminates the polymerisation and creates a chain of twice the length.

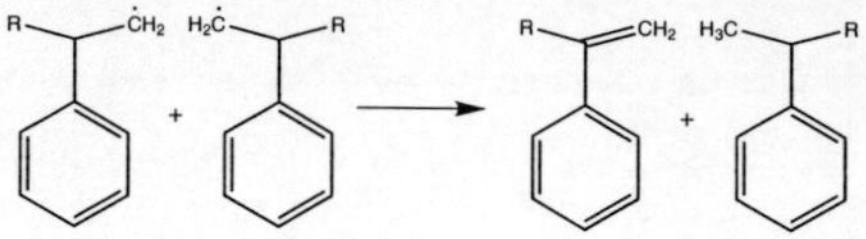

FIG. 4.10: The termination reaction is achieved by disproportionation of two radicals.

and a saturated monomer for the other molecule.

4.1.2   Semi-continuous emulsion polymerisation

An alternative reaction path for a batch emulsion polymerisation is the semi-continuous polymerisation [El-Aasser 90] (see Fig. 4.12). After the initial step of a nucleation phase, also known as seed, a second step is performed. In this second step the reaction proceeds during a permanent inflow of the reactive medium (see Fig. 4.13). The reactive medium is split up in two parts. One containing monomer, SDS, acrylic acid and some water in a pre-emulsion. The second part contains additional water and the initiator. The preparation recipes can be taken from Appendix D. If the inflow of the monomer and initiator is slower than the reaction speed, this process is called a monomer poor reaction. Furthermore, it is possible to synthesise multilayered particles, like core shell systems. Here different monomer or monomer mixtures can be added at different times. This will influence the properties of the particles in the end. Shells or cores with cross-linkers or different T_g are accessible in this way. When the reaction mixture is continuously changed during the reaction [Hoy 79, Lambla 85] particles containing a gradient can be created. This can result in properties which also show this gradient.

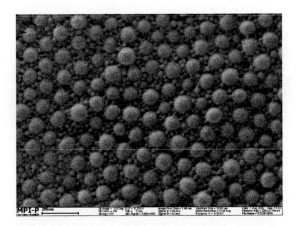

FIG. 4.11: A SEM picture of particles synthesised via semi-continuous emulsion polymerisation. Clearly visible are the small particles that are left over from the seed, whereas the bigger particles are the final product particles.

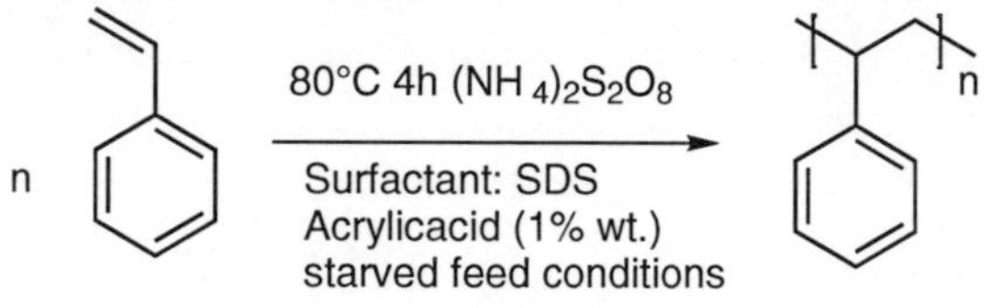

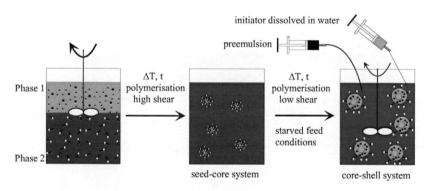

FIG. 4.12: Typical reaction equation and conditions of a semi-continuous emulsion polymerisation.

FIG. 4.13: The three different steps of the semi-continuous emulsion polymerisation are displayed. First a mixture of all ingredients is made to synthesise the seed. Then the seed particles are formed. In a last step these particles grow to the final particles under starved feed conditions. The syringes contain a) the pre-emulsion and b) the initiator dissolved in water.

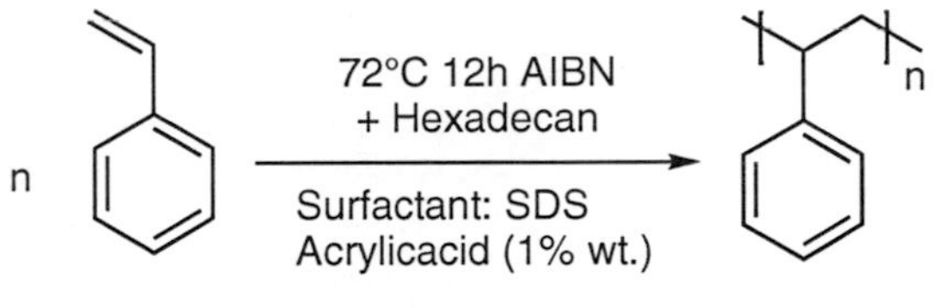

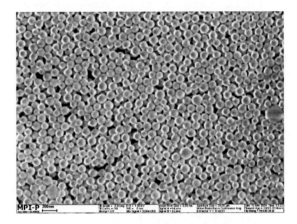

FIG. 4.14: The reaction equation and conditions of the mini-emulsion polymerisation.

FIG. 4.15: A SEM picture of particles synthesised via mini-emulsion polymerisation. The dynamic light scattering gives a polydispersity of 0.01.

4.1.3   Mini-emulsion polymerisation

A further possibility of conducting the emulsion polymerisation is the mini-emulsion polymerisation. These mini-emulsions are stabilised oil droplets with a size between $50\ nm$ and $500\ nm$. They are prepared by high-shear respectively ultra-sonication of a system containing monomer, water, a surfactant and a hydrophobe (see Fig. 4.14). Here the so-called 'nanoreactors' are the centre of the polymer reaction. These nanoreactors, separated from each other by the continuous phase, contain the essential ingredients for the formation of the nanoparticles (see Fig. 4.16). When carefully prepared, polymerisations in such mini-emulsions lead to latex particles of about the same size as the initial droplets. This is due to the fact that the droplets (nanoreactor) do not interact with each other. The

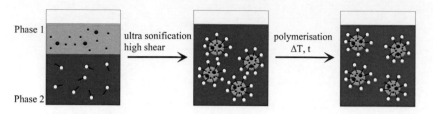

FIG. 4.16: In this sketch the three states of the mini-emulsion polymerisation are visible. In the first state the other reaction ingredients are dissolved in two phases, the organic (phase 1) and the water phase (phase 2). In the second state one can see the micro-compartements after ultra-sonication. The reaction takes place in these mini-emulsion reactors. In the third state the completely polymerised particles are shown.

interaction is suppressed with the addition of a hydrophobe agent and a strong surfactant. The hydrophobe agent stabilises the system against Ostwald ripening or coalescence. The droplets are the main place where the polymer reactions happen. This means that the droplets have to become the primary place of the initiation of the polymer reaction. Then polymerisation proceeds in the droplets, where the water transports new reactive material and heat. Within these 'nanore-actors' the nanoparticles are synthesised. The polymerisation of mini-emulsions extends the possibilities of the widely applied emulsion polymerisation and provides advantages with respect to copolymerisation reactions of monomers with different polarity and incorporation of hydrophobic materials. For the synthesis the emulsion droplets should have the same size (monodispers), and the size should be adjusted. This is done via ultra-sonication, despite the stabilisation of the emulsions by the hydrophobe agent. Then fusion and fission processes are induced, and it can be seen that with increasing ultra-sonication time the size of the droplets can be decreased in a controlled way [Landfester 03]. It is an advantage of this method that in liquid/liquid synthesis, both the direct (aqueous solvent) and inverse (organic or hydrocarbon solvent) situations are possible. Furthermore it can be used to generate nanocomposites with high stability and processability, by the encapsulation of nanoparticles into polymer shells.

4.2   Characterisation methods for dispersions

After the synthesis the dispersions have been characterised. Several methods were used to determine the particle size, the zeta-potential and the solid content. These methods are introduced and presented in this chapter.

4.2.1   Particle size

When an electromagnetic wave travels through a medium, scattering or absorption can occur. The scattering can be elastic or inelastic. In the case of inelastic scatter (e.g. Raman scattering) a shift of the wavelength of the scattered beam is detected. A special case of the inelastic scattering is the quasielasic scattering, where the frequency shift is small compared with inelastic scattering. It appears when electro magnetic radiation interacts with molecules or particles under a translational movement. This frequency shift is referred to as Doppler-effect. Dynamic light scattering (DLS) techniques permit particle size measurements from $5\ nm$ up to $250\ mm$ of particles dispersed in a solvent. Essential for this methods is that the sample does not absorb light, and it has to be diluted that there is no dominant particle interaction and that there is no multiple scattering. Using light scattering, the average size and the size distribution can be determined. For a better understanding of light scattering techniques the reader is referred to literature [Berne 76, Pecora 85, Schramm 90, Wiese 92, Brown 93]. Furthermore some important guidelines have to be fulfilled for the measurements: The scatter of the particles has to be higher than that of the medium. This can be achieved by adjusting the concentration of the particles. On the other hand the concentration should not be too high, because then multiple scattering occurs and the interactions between particles cannot be disregarded. The optical interactions inside a particle can only be neglected when the particle diameter is small compared with the wavelength λ of the incident light beam.

The Brownian molecular motion of colloidal particles results in fluctuation of the scattered intensity, which is acquired by the photo multiplier and then correlated by the correlator. The diffusion constant of particles strongly depends on their size. Smaller particles diffuse faster, thus resulting in a faster fluctuation of intensity. Bigger particles diffuse slower thus resulting in slower fluctuation of intensity. For short time intervals $t_1 = \tau$ und $t_2 = \tau + d\tau$ the intensity fluctuations are small, respectively the positions of the particles are highly correlated with their starting position, whereas in long intervals their positions are just at random. The signal measured with the correlator is the normalised intensity correlation function $g^{(2)}(\tau)$ which is given by [Wiese 92]:

$$g^{(2)}(\tau) = \frac{< I_s(t) \cdot I_s(t + \tau) >}{< I_s(t) >_t^2}. \tag{4.1}$$

Here I_S is the scattering intensity, $\langle ... \rangle$ the time average and τ the correlation time. The spherical particles of the same diameter one can get the dependence of $g^{(2)}(\tau)$ on the diffusion coefficient D out of the equation (4.1) [Wiese 92]:

$$g^{(2)}(\tau) = 1 + \beta \cdot e^{-2Dq^2 \cdot \tau}. \tag{4.2}$$

The experimental constant β can have values between 0 and 1. The absolute value of the scattering vector q is given by:

$$q = \frac{4\pi n_D}{\lambda} sin(\frac{\Theta}{2}), \tag{4.3}$$

here n_D is the refractive index of the medium, λ the wavelength of the incident beam and Θ the scattering angle. Polydisperse spherical particles under free diffusion cannot be described with equation (4.2). The exponential function in equation (4.2) is tended by a sum of exponential functions with different relaxation times $\Xi_i = D_i q^2$ (D_i = diffusion coefficient of particle fraction i) [Wiese 92]:

$$g^{(2)}(\tau) = 1 + \beta[\sum G\Xi_i e^{-\Xi_i\tau}]^2. \tag{4.4}$$

The intensity of the different particle fractions weights the distribution of relaxation times $G(\Xi_i)$. The simplest method of determining the particle size from the correlation function is the cumulant analysis. The expansion of the auto correlation function $ln[g^{(2)}(\tau) - 1]$ is stopped after the third term:

$$\frac{1}{2} ln[\frac{g^{(2)}(\tau) - 1}{\beta}] = -K_1 \cdot \tau + \frac{1}{2}K_2 \cdot \tau^2 - \frac{1}{6}K_3 \cdot \tau^3. \tag{4.5}$$

The coefficients K_1, K_2 und K_3 are called cumulants. The intensity-weighted averaged diffusion coefficient $< D >$ can be calculated via:

$$< D > = \frac{K_1}{q^2}. \tag{4.6}$$

The hydrodynamic radius r_H can afterwards be calculated via the Stokes-Einstein equation (4.7) in case the diffusion coefficient is known:

$$D = \frac{k_B T}{6\pi\eta r_H},\tag{4.7}$$

here k_B is the Boltzmann constant, T is the absolute temperature in K and η is the viscosity of the dispersive medium. The width of the particle size distribution index (PDI) is derived from the second and the third cumulant:

$$PDI = \frac{K_2}{K_1^2}.\tag{4.8}$$

It is a disadvantage of the cumulants method that on cannot distinguish between a broad or a bimodal distribution. Therefore, one gets only an average particle diameter and a value of the width of the distribution. Alternatively the Contin algorithm could be used to analyse $g^{(2)}(\tau)$ [Provencher 82a, Provencher 82b]. The Contin algorithm is able to distinguish between a broad or a bimodal distribution. The hydrodynamic radius r_H, and the polydispersity are then calculated via the selected method.

For the determination of the particle size and the zeta-potential (see chapter 4.2.2) a Malvern Zetasizer 5000 is used. The incorporated Correlator is of a type 7132 with 62 correlation channel. The light source is a 5 mW HeNe-LASER at a wave length of $\lambda = 633\ nm$. In these machines photon correlation spectroscopy (PCS) is used. In this particle sizer the scattered light is detected at an angle of 90 °. For the measurements the dispersion are diluted with 'Millipore water'. The resulting solid content must not be bigger than 0.1 $wt.\%$. This amount was filled in the measurement cuvette and then analysed.

4.2.2 Zeta-potential

The stability of colloidal solutions strongly depends on the charges at the particle-liquid interface [Dörfler 02, Hunter 88]. Because of net negative charges on the particle surface, particles repel each other and flocculation does not occur. Most important sources of charges are the ionisation of chemical groups at the particle surface and the ability of the solution to absorb the differently charged ions. A net charge at the particle surface results in an increased counter ion concentration. The structure of a charged surface and surrounding counterion charges is called an electrical double layer. The outer layer consists of two different parts, an inner one containing relatively strong bound charges and an outer diffuse layer where the thermal motion and electronic forces balance each other. This electric potential decreases and decays with the increasing distance to the particle surface. In the bulk the potential is totally screened by the surrounding ions. The zeta-potential occurs at the surface and is build of the particle and the inner electric layer. Electrophoresis is one way to measure this potential. That means, the movement of charged particles, which is induced by an electric field, is analysed in a liquid. While applying an electric field, the charged particles are attracted to the opposite charged electrode. This movement is opposed by the viscosity of the surrounding media. After reaching an equilibrium the constant velocity depends on the electric field strength E, the dielectric constant ϵ, the viscosity of the medium η and the zeta-potential ζ. The electrophoretic mobility U_E of a particle is defined as the ratio of the measured velocity ν, and the applied electric field E. U_E is related to the zeta-potential by the Henry-equation [Hunter 88]:

$$\frac{\nu}{E} = U_E = \frac{2\epsilon\zeta}{3\eta}f(\kappa a). \tag{4.9}$$

The term $f(\kappa a)$ is called the Henry-function:

$$f(\kappa a) = 1.5 - \frac{9}{2(\kappa a)} + \frac{75}{2(\kappa a)^2} - \frac{330}{(\kappa a)^3}. \tag{4.10}$$

This function depends on the shape of the particles. For a spherical particle and a large κa it approaches a value of 1.5. This approximation is called the Smolu-chowski limit, which is valid for particles of about 0.1 microns and an electrolyte with more than $10^{-3} \frac{mol}{l}$ of salt and results for equation (4.9) into:

$$U_E = \frac{\epsilon\zeta}{\eta}. \tag{4.11}$$

The mobility is measured via LASER Doppler velocimetry (LDV). Here two LASER-beams are crossed. At the crossing an interference pattern is created. Due to the electric field, particles are moving. The resulting intensity fluctuations are digitised and correlated. With the analysis of the correlation spectra the mobility of the particles and therefore the zeta-potential is determined.

4.2.3 Scanning Electron Microscopy

Normal optical or ultra microscopy does not achieve the spatial resolution needed for the imaging of the particles. Particles with a size down to $20\ nm$ are visualised with scanning electron microscopy. More detailed information can be found in the literature [Bindell 92, Goldstein 92, Sawyer 96]. This method provides a good possibility of visualising at these small structures in a relatively short time. The SEM is based on an electron beam creating core, consisting of an electron gun and magnetic lenses. The electrons, produced by the gun, have energies between 0.2 and $30\ keV$. The lenses reduce the electron beam in diameter and focus it on the sample in a vacuum. Magnetic lenses move the electron beam over the surface of the sample in a scanning pattern. The reflected electrons and the emitted photons carry the information of the surface of the sample. Detectors for secondary (SE) and backscattered electrons (BSE) collect the electrons. The high energetic BSE (energy higher than $50\ eV$) have their origin in the deeper layers of the sample. Their contrast depends on the topography and the average atomic number of the sample. The low energy electrons have their origin in the surface layers of the sample and they transport mainly topographic information. If a modern SEM is used, a small gold layer on the surface of the samples is not necessary, this is due to small energy of the electrons. In older set-ups samples might be damaged [Butler 95, Vezie 95, Jaksch 95, Joy 96] by the electron beam. If there is a need to go to smaller a resolution than $10\ nm$ with the low acceleration voltage, then again a gold layer is necessary.

The measurements within this thesis have been performed on a low voltage-SEM (LV-SEM) of the company LEO (Type Gemini 1530) with a SE-detector. The acceleration voltage can be varied between 0.2 and $3\ keV$. It is primarily used for the examination of the size of the particles and the particles size distribution. The dispersion is dropped on a silicon wafer and then dried. The so prepared sample is then directly examined in the SEM.

4.2.4 Experimental issues of the synthesis

In this chapter the experimental issues of the synthesis and the characterisation are presented. Typical procedures for the synthesis can be taken from Appendix D. First two mixtures in separate beakers are made. The first contains water, surfactant initiator, and the second containing monomer, hydrophobe, acrylic acid and co-monomer. After mixing these two beakers and stirring it for one hour, the solution is sonicated for 2 minutes. The reactive medium is cooled during the sonication. The maximum elongation that is allowed for this sonifier tip was used. In this case it has a value of 89 %. The used sonifier is a Branson Digital Sonifier ®Model 250-D with a resonator $\frac{1}{2}$" (W-250) of 200 W. Afterwards the mixture is taken to the reactor and the synthesis is performed. For getting the temperature constant a Huber Ministat was used.

Directly after the synthesis the pH-value was set to 10, where the dispersion are stable over a time span of at least 1 year. The pH-value of the dispersions was set with a WTW pH 320 Set-2. This pH-meter was calibrated with buffers at a pH-value of 2, 7 and 10.

The solid content of the dispersions is determined via gravimetric analysis. Furthermore, the conversion of the synthesis can also be determined via this method. The solid content is commonly defined as the mass of the dispersion after drying. The solid content comprises the polymerised monomer, the surfactant, and non-volatile components of the initiator. The mass of the initiator due to its small amount is negligible. The initial mass is m_1. The residual mass is m_2. The solid content is then calculated:

$$solid\ content = \frac{m_2}{m_1}. \tag{4.12}$$

For the analysis of the solid content a small amount of the dispersion, normally about 2 g, is put in a small aluminium jacket and the volatile parts are evaporated under vacuum. The dispersion are heated for 12 hours at 60 $°C$ in a vacuum of below 100 $mbar$.

Chapter 5

Materials

5.1 Dispersions

The dispersions are synthesised via emulsion polymerisation. The two different pathways of the synthesis are semi-continuous emulsion polymerisation and mini-emulsion polymerisation. These two different methods are presented in chapter 4. Both are based on radical polymerisation, as described in chapter 4.1.1. The Debye-length of the dispersions is calculated according to equation (2.2.1). The Debye length of the dispersions synthesised via semi-continuous emulsion polymerisation is in the range of $0.3\ nm - 0.4\ nm$, whereas the Debye-length of the dispersions synthesised via mini-emulsion polymerisation is in the range of $1.7\ nm - 1.8\ nm$.

5.2 FD-Virus

The rod shaped semi-flexible bacteriophage FD-virus [Fraden 95] is a polyelectrolyte that changes its state with increasing concentration from an isotropic solution to a cholesteric liquid crystal. The FD-virus has a contour length of $880\ nm$, a bare diameter of $6.6\ nm$ and a persistence length of $2,200\ nm$. The rotational diffusion coefficient D_r is $20.9\ s^{-1}$ [Newman 77]. The overlap concentration c^* is $0.04\ \frac{mg}{ml}$. The stock solution of FD-virus ($12\ \frac{mg}{ml}$) used for these experiments was kindly supplied by the group of Prof. J. Dhont from the Forschungszentrum Jülich. The stock solution is prepared via the method described by Marvin et al. [Marvin 75]. The different concentrations are obtained by diluting the stock so-

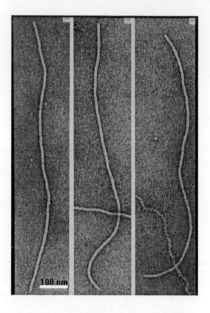

FIG. 5.1: SEM picture of a FD-virus.

lution with a $20\ mM$ tris(hydroxy methyl)-aminomethan chloride (Tris-Cl) buffer solution. The effective diameter changes with the salt concentration of the stock solution. In this case it is $20\ nm$. The precise concentration is determined by ultraviolet spectroscopy (UV) absorption at a wavelength of $270\ nm$, using a specific absorption coefficient of $3,84_{(1cm,269nm)}\ \frac{mg}{ml}$ [Day 88]. Twelve microliter of the virus stock solution are diluted with one milliliter of a $20\ mM$ Tris-Cl buffer solution. A UV-spectra is then recorded. The concentration c was then calculated via equation (5.1). The quality of the solution can also be determined via UV-spectra (see equation (5.2)).

$$c = \frac{OD_{270nm}}{3.84\frac{l}{mol}}\frac{1ml}{1.12ml} \cdot 100, \tag{5.1}$$

$$\frac{OD_{270nm}}{OD_{240nm}} \approx 1,4 \pm 0,1. \tag{5.2}$$

The FD-Virus solution is stored at $4\ °C$ to prevent growth of bacteria. For cleaning the solution from bacteria after measurement the solution is centrifuged for 10 minutes at about $3,000\ g$. For the reconcentration of the solution after use, an ultracentrifugation for $5\,h$ at a temperature of $4\,°C$ and $100,000\,g$ is necessary. From this concentrated solution a new dilution series can be started. For centrifugation a Heraeus Megafuge 1.0, and Beckman L8-M Ultracentrifuge with a SW28 rotor have been used.

5.3 Polystyrene in DOP

The high-molecular-weight PS sample is synthesised via anionic polymerisation ($M_n = 2.6 * 10^{-6}\,\frac{g}{mol}$, $\frac{M_w}{M_n} = 1.3$). The PS sample is dissolved in dioctyl phthalate (DOP) at a concentration of $8\,wt.\%$ using a co-solvent methylene chloride [Neidhöfer 03a]. This concentration corresponds to an effective entanglement molecular weight of $450,000\,\frac{g}{mol}$ due to the $M_n \propto \varphi^{\frac{3}{4}}$ dependence [Colby 90], and equals an effective number of entanglements of 5.8. After loading the sample into an optical Couette cell or optical parallel plates, it is left at rest for 20 minutes, to assure complete relaxation of the material. A similar system was already investigated in rheo-optical studies [Hilliou 02] and used as a reference material.

Chapter 6

Improvement of a rheo-optical set-up

Optical methods are commonly used because of their sensitivity, dynamic range and the micro structural information that they can provide. Rheo-optical measurements as an in situ method are used to analyse the dynamic and structural properties in general and of colloidal and polymeric samples [Arendt 98, Clasen 01, Fuller 90, Janeschitz-Kr 83, Kulicke 98, Kumaraswamy 99, Lodge 94, Mewis 97, Peterlin 76, Ven 90, Linden 03, Wagner 98] specifically under an external mechanical field. By a simultaneous detection of mechanical and the optical signals birefringence and dichroism, it is possible to correlate the macroscopic mechanical response and the microscopic induced changes in the material. Because the commercially available set-up, was not sensitive and fast enough the improvements of the experimental set-up to enabled to measure with higher sensitivity, time resolution and to use it also for dynamic measurements at higher frequencies. Emphasise is placed on the hard and software related reduction of the background birefringence present in the system. Finally several experimental results will be presented to quantify the achieved several fold improvements.

6.1 Theory of birefringence and dichroism

In any rheo-optical set-up the polarisation and the amplitude of the electromagnetic light is analysed. The dependence of the optical properties, e.g. of the anisotropy of the refractive as induced via flow, stress or deformation is determined. For understanding the principle of rheo-optics one has to be aware of the nature of the light and of the refractive index. The oscillating electric field $\underline{E}$ of the light equation (6.1) is influenced by the electrons and nuclei of the material is passes through.

$$\underline{E} = \underline{E}_0 cos(\frac{2\pi \underline{n}}{\lambda}z - \omega t) \tag{6.1}$$

The polarisability of this material causes the light beam to alter its velocity. This effect is macroscopically known as the refractive index n. In anisotropic material (e.g. as caused by flow, deformation or crystal structure) the polarisability and therefore the refractive index (both anisotropic) are be represented by a complex tensor equation (6.2).

$$\underline{n} = \underline{n}' + i\underline{n}'' \tag{6.2}$$

In this representation the real part $\underline{n}'$ describes the phase shift, and the imaginary part $\underline{n}''$ describes the extinction (or attenuation) of the amplitude [Kerker 69]. The trace of the tensor $\underline{n}$ is reflected to as the refractive index. In the direction of propagation of the light z, the component $n'_{z,z}$ of $\underline{n}'$ and the component $n''_{z,z}$ of $\underline{n}''$ are zero. The difference of the components $n'_{x,x}$ and $n'_{y,y}$ is called birefringence $\Delta n'$ and the difference of the components $n''_{x,x}$ and $n''_{y,y}$ is called dichroism $\Delta n''$. The components of the electric field vector in the two directions of the plane x, y perpendicular to the direction of propagation of the light z can then be written as the following:

$$\underline{E} = \begin{pmatrix} E_{0x}cos(\frac{2\pi \Delta n'_{x,x}dz}{\lambda} - \omega t) \\ E_{0y}cos(\frac{2\pi \Delta n'_{y,y}dz}{\lambda} - \omega t) \end{pmatrix} \tag{6.3}$$

here

$$\delta' = \frac{2\pi \Delta n'd}{\lambda} \tag{6.4}$$

is called the retardation δ', d the optical path length, λ the wavelength of the light source and $\Delta n' = n'_{x,x} - n'_{y,y}$ the birefringence. The orientation angle θ is defined by:

$$tan2\theta = \frac{\Delta n'}{\Delta n''} = \frac{n'}{n''} \tag{6.5}$$

The correlation between the applied mechanical field and its effect on the birefringence is given by equation (6.6). Here are the σ stress tensor and the C the stress optical coefficient , which is unique for every material [Fuller 95, Larson 99].

$$\underline{n} = C\underline{\sigma} \tag{6.6}$$

The equation (6.6) can be reduced to the following expression with shear stress σ_{12} change in birefringence $\Delta n'$ and orientation angle θ accessible via measurements.

$$\sigma_{12} = \frac{1}{2C}\Delta n' sin2\theta. \tag{6.7}$$

In the following section the determination of these quantities will be described in detail.

6.2 Rheo-Optical set-up

6.2.1 Experimental set-up

All experiments were performed on two different ARES rheometer (Rheometrics Scientific) both equipped with a commercial Optical Analysis Module OAM supplied by Rheometrics Scientific. The ARES instrument is a strain-controlled rheometer. Both instruments were equipped with the standard motor and with a dual range Force Rebalance Transducer (2K FRTN1) ranging from $2 * 10^{-6}$ Nm to $2 * 10^{-1}$ Nm. All experiments were carried out using a commercial and a home-built optical Couette flow cell. In case of the home built unit the static inner bob (built of brass) has a diameter of 30 mm, whereas the rotating outer cup has an inner diameter of 33.8 mm and is equipped with a quartz bottom plate. The optical unit consists of a solid-state laser (670 nm, 5 mW), a polariser cube, a half-wave plate spinning around at approximately $\frac{\omega_1}{2\pi} =$ 400 Hz and a beam splitter (Fig. 6.1). As a re-

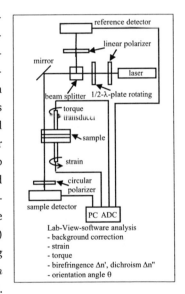

FIG. 6.1: Schematic descriptions of the rheo-optical set-up, including optical train and data analysis in the attached PC.

sult of the rotating half-wave plate the resulting signal is modulated at 1,600 Hz $\pm$ 5 Hz. The variation of $\pm$ 5 Hz is due to mechanical and electronic instabilities. The LASER beam is split and one beam is directed through a linear polariser to a photo diode generating the reference signal. The other part of the beam is directed through the sample to a second photodiode. In case dichroism is negligible a circular polariser is placed in front of the sample photo diode. A circular polariser consists of a linear polariser at $0\,°$ glued together with a half-wave plate at $+45\,°$. This way it is possible to measure directly the birefringence. The relevant equations for the measured intensity signal of the sample equation (6.8) and the reference equation (6.9) can be derived via the Mueller matrices formalism (see also Appendix A.1 on page 181 and [Fuller 95]):

$$I_S = I_{0,S}(1 + sin\delta' cos(4\omega_1 t(\theta - \frac{\pi}{4})))$$
(6.8)

$$I_R = I_{0,R}(1 + cos(4\omega_1 t))$$
(6.9)

with

$$4\omega_1/2\pi = 1,600 Hz$$
(6.10)

where:

I_S: Signal intensity at the sample detector,

I_R: Signal intensity at the reference detector,

$I_{0,S}$: DC signal intensity at the sample detector,

$I_{0,R}$: DC signal intensity at the reference detector.

Here the retardation δ' (see also equation (6.4)) is the retardation of the light after traveling through the sample, including the two parallel quartz plates. By analysing the intensity of the detected light of the reference and the sample beam at a frequency of $1,600\ Hz$ the value of δ' and orientation angle θ can be determined (note: δ'' is called extinction in the case of dichroism). The birefringence $\Delta n'$ can be calculated from the retardation by using equation (6.4).

6.2.2   Commercial set-up

In the commercial set-up a lock-in amplifier analyses the optical signals. A 12-bit ADC is used to digitise the signal of the lock-in amplifier. This ADC limits the dynamic range of the set-up to $1 : 4,096$. Subsequently a separate software module calculates the dichroism and/or the birefringence together with the orientation angle. The software additionally limits the fastest sampling rate for these observables to be $10 \ s^{-1}$. Furthermore the set-up does not provide a triggering signal to facilitate simultaneous dynamic measurements of the rheological and optical signals. In addition to these problems the optical unit is triggered by the rheometer for step rate and step shear experiments only. Using a lock-in amplifier has the disadvantage that it is fixed on a specific frequency, and therefore it does not follow the fluctuations of $\pm \ 5 \ Hz$. A new solution for this problem is described in the next chapter.

6.2.3 Modified set-up

The here presented set-up improved the original hardware, respectively the original optical train but includes several modifications. For the further data treatment the raw signals were read out on the BNC connections at the back of the Optical Analysis Module (see Fig. 6.1). The analog optical signals where digitised by a 16-bit ADC (the dynamic range is increased to $1 : 65,536$), using a PCI-MIO-16XE 10 (National Instruments, USA), which has a maximum sampling rate of $100 \ kHz$ and is capable of multiplexing up to 16 channels. The used ADC's can simultaneously acquire and transfer the data to the PC memory by data-buffering techniques. As a consequence the optical and rheological data is intrinsically synchronised. The $40 \ \mu s$ interchannel delay (time between consecutive data points) between the four channels is relatively insignificant compared to the timescale of rheological experiments. For the stress and strain signals a simple averaging is conducted in the time domain on-the-fly [Dusschoten 01]. To be able to measure a high resolution in the time domain, especially to detect frequencies up to 12 Hz in dynamic measurements, one needs to sample at least at $24 \ Hz$ according to the Nyquist theorem. Since it is possible to acquire 25 blocks of $1,000$ data points per second ($25,000$ data points per second and channel) a discrete Fourier transformation after several minutes of acquisition would require too much time for the computer to keep up with the ADC. Therefore the signal is divided into blocks of $1,024$ points and a Fast Fourier Transformation (FFT) is done. To form this specific FFT algorithm one needs 2^n points for calculation. The optical and rheological data is then averaged over about $40 \ ms$ ($1/24,576 \ s^{-1}$). As has been explained by Wilhelm et al. [Wilhelm 98, Wilhelm 99, Wilhelm 02] this means that Fourier peaks will arise exactly at the frequency of which the shear oscilla-tions is conducted due to the δ-peak character in frequency space of the mechan-ical excitation. The ADC's sampling rate to four times $24,576 \, Hz$, which means each channel is effectively scanned by $24 \, Hz$ after taking the average of $1,024$ data points. By reducing the data blocks to 512 data points the maximum overall scanning rate is about $48 \, Hz$. This gives an increase in the signal to noise ratio of the phase angle determination of the optical signal (see 6.2). A comparison of the different noise level depending on the number of points for the FFT is shown in Fig. 6.2.

The time dependent data was collected and further processed via a home writ-ten LabVIEW program see chapter A. In this application 4 channels (strain, torque, reference, sample) are used, each at a sampling rate of $24,576 \, Hz$. A descrip-tion on the handling of the rheological signals (strain and torque) is found in [Dusschoten 01] and limit ourselves to the treatment of the two optical signals. The optical data is collected by 2 of the 4 ADC channels in blocks of 2^n points and subsequently an on the fly a Fast-Fourier Transformation (FFT) is performed. Using the oversampling technique a several fold-improved sensitivity is achieved

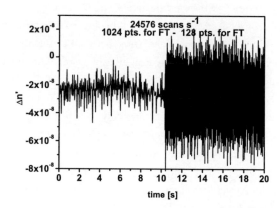

FIG. 6.2: Comparing noise levels of the acquired optical signals in the modified set-up with a sampling rate of $1/24,576 \, s^{-1}$, but different numbers of oversampling data points. Clearly the increase from 128 to $1,024$ reduces drastically the noise level.

and a variable lock-in amplifier is emulated by calculating the Fourier spectrum and follow online the fluctuations of $\pm 5\,Hz$. Using the FFT a frequency spectrum for both reference and sample signal are generated online. The peaks of interest are the one around $1,600\,Hz$ and the $0\,Hz$ peak, the last one corresponding to the DC component, which is equivalent to the offset. Since the angular speed of the half-wave plate is not constant ($1,600\,Hz \pm 5\,Hz$) the modulation frequency is fluctuating over time and is constantly monitored. The reference beam is analysed typically 24 times per second to determine the momentary modulation frequency and is therefore leading to improved performance as compared to the standard set-up using to a fixed lock-in amplifier at a constant frequency. At the online detected modulation frequency the data is taken from the sample frequency spectrum. In this way it is assumed that the actual modulation frequency is measured. To probe $1,600\,Hz$ the sampling has to be done with at least twice the frequency ($3,200\,Hz$) to comply with the Nyquist theorem. Using a tunable hardware active low pass filter, the absence of aliasing from higher frequencies is checked. Typically blocks of $1,024$ (2^{10}) points are used prior to the FFT. In this way about 60 periods with 15 points per period are sampled prior to the FT calculation. This oversampling increases the S/N ratio of stochastic noise by up to a factor of $60^{1/2}$ ($= 7.71$) relative to the situation without having analysed a single period of the time signal. Furthermore by increasing the number of periods the spectral resolution in the frequency spectrum is increased. This procedure results in an effective sampling rate of the birefringence/dichroism of $24\,Hz$. The further data treatment of the individual data points of the optical data, originating from the reference and sample channels, require a significant amount of processing in order to obtain the birefringence $\Delta n'$ and orientation angle θ. The following steps have to be performed:

1. Calculation of the phase of the birefringence from the sample signal relative to the reference signal.

2. Normalisation of the peak at $4_{\omega_1 t}$: $\frac{I_{(S,4\omega_1)}/I_{(S,DC)}}{I_{(R,4\omega_1)}/I_{(R,DC)}}$.

3. Calculation of $\Delta n'$ and θ, see equation (6.8).

4. Finally the time dependent birefringent data is then plotted.

6.2.4   Homebuilt bob for the optical Couette cell

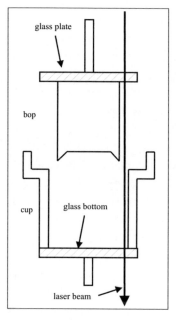

FIG. 6.3: Homebuilt bob with glass plate to avoid overflow, cup with glass bottom.

In the commercial set-up the bob has a small quartz glass plate window at the point where the LASER beam enters the sample from the top. This glass plate window has a diameter of only 1 cm and a ring of stainless steel surrounds it. This set-up is useful for samples with low viscosity, but in samples with higher viscosity, respective normal forces, secondary flows are present and the sample may run over the glass plate. These inhomogeneities have obviously negative effects on both the rheological and optical measurements. To prevent this overflow via secondary flows a new unique bob-design was invented (see Fig. 6.3). On top of the inner shear cylinder a quartz glass circular plate was placed, so that the whole surface of the sample above the gap is covered. On the shaft of the inner bob the glass plate is placed to assure a flat surface of the sample at the point where the laser beam enters the material. For the alignment of the optical cell prior gluing, inserted a close fitting cylindrical core between the bob and cup and concentrically fixed the whole set-up in the rheometer while the glue was curing. This minimised changes in the gap size during the experiment. By imposing a normal force and probing time dependence during the curing of the glue a good connection between glass plate and cup was ensured. It was checked that the quartz glass had no influence on the optical signal by rotating the empty cup. Consequently it was assumed that the flow cell did not dominantly cause the detected overall birefringence of the background signal.

6.2.5 Background subtraction by hard and software measures, ambiguity of the orientation angle and calibration

Due to residual birefringence within the optical train (i.e. caused by the prisms and misalignments within the optical train) the background birefringence is a non-zero vectorial quantity. This vectorial quantity will add to the measured optical signal and especially in the situation where the sample signal is of the order of the background signal, large deviations will occur between the measured birefringence signal and the birefringence signal as caused by the sample (see Fig. 6.4 in case of clockwise and Fig. 6.5 in case of counterclockwise rotation). These deviations occur most pronounced in the measurements of the orientation angle at small shear rates (see Fig. 6.6). Since one can assign a magnitude (the retardation) as well as a direction (the orientation angle) to a birefringence signal it is possible to represent the birefringence as a vector. First using a variable retarder in front of the sample detector minimised by hardware the background birefringence. Here two birefringent objects with different orientations were put at a certain angle relative to each other and they intercept at 180 ° relative to the instrumental birefringent vector. Consequently a vectorial compensation of the background birefringence is achieved (see Fig. 6.7).

Although this reduced the background retardation to typically less than 1 % of the original signal, still the presence of this small leftover background will remain and cause interference at small enough shear rates. Especially the measurements of the orientation angle are sensitive towards small background signals (as shown in Fig. 6.6) because the orientation of the resulting vector still strongly depends on the direction of the background birefringence. In order to compensate the influence of the background signal, two retarding elements were considered in series, the first (corresponding to the measured signal) with retardation o and orientation angle ϵ and the second (corresponding to the background signal) with retardation p and orientation angle γ. As shown in Appendix A.1 on page 181 using the Mueller matrix formalism in cases where the retardation is small enough (so that may be used $sin\delta' \approx \delta'$) the following relation holds for the resulting signal (corresponds to the true birefringence signal) with retardation q and orientation angle φ:

$$q^{(i2\varphi)} = o^{(i2\epsilon)} + p^{(i2\gamma)} \tag{6.11}$$

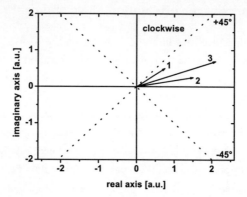

FIG. 6.4: Vector diagram in case of clockwise rotation. The bolt drawn vector 3 should represent the measured birefringence including the background vectorial birefringence. The short vector 1 represents the background. The third vector 2 shows the sample's birefringence and is calculated via subtracting the background vector 1 from the measured vector 3. Vector diagram in case of counterclockwise rotation.

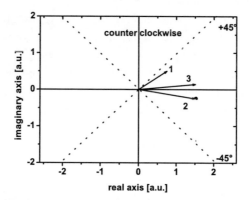

FIG. 6.5: Vector diagram in case of counterclockwise rotation. The bolt drawn vector 3 is the measured birefringence including background. The short vector 1 represents the background mirrored at the y-axis. The third vector 2 shows the sample's birefringence and is calculated via subtracting the background vector 1 and the measured vector 3.

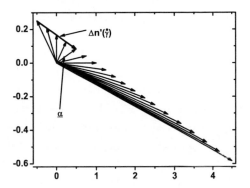

FIG. 6.6: Typical set of measured vectors for a counterclockwise rotation. Vector $\underline{\alpha}$ represents the minimum of the measured birefringence value see Fig. 6.14. $\Delta n'(\dot{\gamma})$ represents the measured birefringence for an oscillating shear rate.

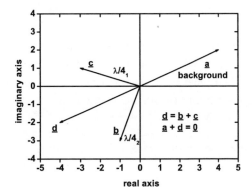

FIG. 6.7: Vectorial representation of the compensation of the birefringence background. Vector $\underline{a}$ is the original background. Vectors $\underline{b}$ and $\underline{c}$ represent the birefringence of the two $\frac{\lambda}{4}$-plates of the variable retarder. Vector $\underline{d}$ is the resulting birefringence of the variable retarder that counteracts the background birefringence.

With this numerical calculation the leftover residual birefringence, done by using the variable retarder, will be compensated. The overall birefringence of several materials adds like vectors in the complex plane with a magnitude equal to the retardation and angles orientation doubled. Using this relation one can counter-act using the two retarding elements the background from the measured signal (for a more generalised expression for the reduction of the background reduction see Appendix A.2 on page 182). Since the orientation angles are always only uniquely mapped in the range between $-\frac{\pi}{4}$ and $\frac{\pi}{4}$, the background signal is, how-ever, ambiguous. If the measured angle of the background is e.g. $\theta_{background}$ due to the ambiguity the angle could also be $\theta_{background} + k\,\frac{\pi}{2}$ with k an integer. When subtracting the background from the measured signal one can do it either taking the background with angle $\theta_{background}$ or the background with $\theta_{background} + k\,\frac{\pi}{2}$. This ambiguity can be removed by calibrating the angle θ symmetrical around the direction of flow during clockwise and counter-clockwise rotation of the cup. This calibration of the orientation angle is done by placing a quarter-wave plate with the so-called fast axis (optical axis with low refractive index, perpendicular to the slow axis with a higher index of refraction) in the direction of flow. Alternatively, a sheet polariser with the main axis at $+ 45$ ° relative to the flow direction could be used. This follows from the general expression for the intensity as a function of reorientation at the sample detector (see Appendix A.3 on page 183). For any ex-periment the following calibration procedure, including hardware compensation of the birefringence and definition of the zero angle θ along the tangential of the Couette cell is suggested. This calibration procedure consists of the 4 following steps:

1. Minimising the residual background using a variable retarder made of two birefringent objects, so that $sin\delta' \approx \delta'$ is fulfilled (e.g. $\delta << 20$ °).

2. Calibration of the zero orientation angle by using a polariser (positioned at 45 °) or a quarter-wave plate (at 0 °).

3. Determination of the magnitude of the residual background of the system under zero shear conditions.

4. Checking the efficiency of the hardware birefringence compensation of the background by turning the sample clockwise and counter clockwise. If no

difference in the measured signal the background signal was detected a sufficient compensation was achieved.

Note, that once the background has been measured, minimised, and again quantified the ambiguity of the background signal has been removed. For successive experiments one only needs to perform steps $3 + 4$ as long as the condition $sin\delta' \approx \delta'$ is fulfilled at zero shear.

6.3 Background birefringence and experiments

In the experiments the samples polystyrene solutions (PS) and FD-virus dispersions were investigated. In an ideal case for the retardation δ' the relation $sin\delta'$ $= 1$, $\delta' = 90$ ° applies irrespective of the orientation of the polariser. For a rotating quarter-wave plate without a background signal $sin\delta' = 1$ is expected, see Fig. 6.8, Fig. 6.10 and Fig. 6.12. After minimising the background signal using two $\frac{\lambda}{4}$-plates, see Fig. 6.7, a linear polariser was placed on top of the cup and rotated at the position of the sample to calibrate the angle θ and validate the accuracy of this birefringence measurement. In addition the signal of a rotating quarter-wave plate is measured, see Fig. 6.8, Fig. 6.10 and Fig. 6.12, which showed to be quantitatively the same behaviour as in the simulation in Fig. 6.11. As already mentioned, the background birefringence has a large influence on the birefringence measurements. To illustrate this influence the behaviour of a turning birefringent object (quarter wave plate) with a small and large background was simulated see Fig. 6.9, Fig. 6.11 and Fig. 6.13. The simulation is visualised as a vector diagram in Fig. 6.9. The length of the vectors represents the $sin\delta'$ (retardation), which is the relative change of the magnitudes of sample and reference signal. The angle between the x-axis and the vector gives the orientation angle.

For measurements at higher shear rates, longer optical path lengths or with highly birefringent materials, $\Delta n'$ might deviate from the linear dependence see equation (6.7) of the stress optical coefficient as a function of shear rate. This non-linear dependence can be followed in the shear rate dependent $\Delta n'$ signal (see Fig. 6.14). One can find the maximum detectable $\Delta n'$ during the calibration process, when a quarter-wave plate is inserted into the beam. The maximum value for $sin\delta' = 1$ is then detected and one can easily calculate via equation (6.3) the corresponding maximum birefringence. For the described set-up and the FD-virus $\Delta n'_{max}$ is approximately of $8 * 10^{-6}$. With a highly concentrated ($c = 8$ $\frac{mg}{ml}$) sample of FD-virus it was possible to monitor several orders of birefringence (Fig. 6.14 and Fig. 6.15). To circumvent this problem one could use either smaller shear rates, lower concentration or short optical path length. In Fig. 6.16 and Fig. 6.17 one can see results of a sweep in shear rates starting from $\dot{\gamma} = 1 \ s^{-1}$ and up to $\dot{\gamma} = 250 \ s^{-1}$ for FD-virus at a concentration of $40 \ c^*$. Measuring for $20 \ s$ at the steady shear and then averaging the birefringence data acquired a precise value for $\Delta n'$ for every applied shear rate could be detected. The slope of the shear rate dependent birefringence in Fig. 6.18 is due to equation (6.7) expected to be 1. A value of $m = 0.83$ is found. This might be explained that the function $\sigma \propto \Delta n'$ is not anymore valid due to optical non-linearities. In Fig. 6.17 one can see the data of the corresponding orientation angle. The two branches derive from clockwise and counterclockwise measurements. To validate the accuracy of the measurements for both shear directions the data with a positive sign was multiplied with minus one and then plotted as a line. The line and the data points from the other measurement overlay very well. For a diluted FD-virus sample, with a concentration of $0.06 \ \frac{mg}{ml}$, $\Delta n'_{min}$ of $1 * 10^{-8}$ could be achieved (see Fig. 6.18).

In Fig. 6.20 and Fig. 6.19 the measured birefringence and orientation angle values acquired by the home written LabVIEW software (see Appendix B.3) are compared with the original Rheometrics software. To compare the two set-up's, measurements were performed on standard system [Hilliou 02]. An $8 \ wt. \ \%$ polystyrene ($M_n = 2.6 * 10^{-6} \ \frac{g}{mol}$) solution is used for these experiments. Experiments in steady shear with start and cessation of flow were performed. The applied shear rate is $20 \ s^{-1}$. The time resolution of the LabVIEW data is significantly improved by the factor of 24 and the sign of the orientation angle is correct, see Fig. 6.19 and Fig. 6.20 [Hilliou 02]. The Rheometrics Software determines a positive birefringence for PS-DOP while it is known that a negative sign is correct. Compared to the literature values the birefringence values as computed by the Rheometrics Software are at least 10 times too big [Hilliou 02]. While using this modified set-up an overshoot in the optical signal is monitored (see Fig. 6.20 and Fig. 6.19), whereas it is not visible using the commercial software. Furthermore the correct [Hilliou 02] stress optical coefficient for a $8 \ wt.\%$ PS in DOP solution ($C = 4 * 10^{-9} \frac{m^2}{N}$ compared to the literature values $C = 5 * 10^{-9} \frac{m^2}{N}$) is computed. Until now the focus was on steady shear measurements only. In dynamic measurements the background birefringence has obviously a similar influence. In addition clockwise and counterclockwise movements are conducted in quick succession. As a consequence the correction has to be done alternating for every shear direction. Due to the LASER drift in λ, it is necessary to check and recalibrate the background compensation after every measurement. The background birefringence will modulate via the applied oscillatory shear $\Delta n'$ and can therefore strongly influence the results.

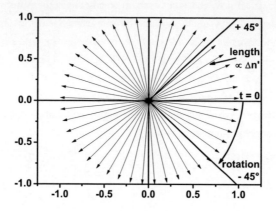

FIG. 6.8: Vectorial representation of the birefringence signal of a turning polariser without any background birefringence.

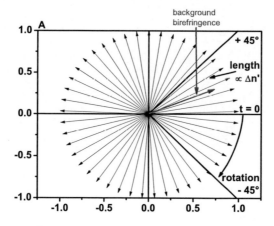

FIG. 6.9: Vectorial representation of the birefringence signal of a turning polariser with background birefringence.

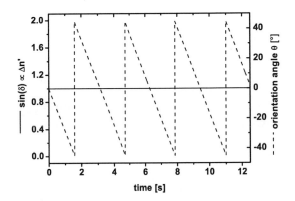

FIG. 6.10: Resulting plot from Fig. 6.8. The length of the vector representing the birefringence $\Delta n'$ and the angle corresponds to the orientation angle θ.

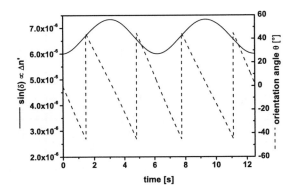

FIG. 6.11: Resulting plot from Fig. 6.9. The length of the vector representing the birefringence $\Delta n'$ and the angle corresponds to the orientation angle θ.

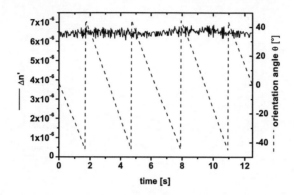

FIG. 6.12: Birefringence $\Delta n'$ and orientation angle θ of a rotating $\frac{\lambda}{4}$-plate with background correction. The frequency is $\frac{\omega_1}{2\pi} = 0.16\ Hz$.

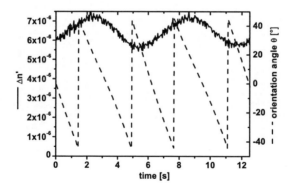

FIG. 6.13: Birefringence $\Delta n'$ and orientation angle θ of a rotating $\frac{\lambda}{4}$-plate without background correction. The frequency is $\frac{\omega_1}{2\pi} = 0.16\ Hz$.

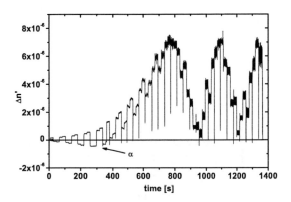

FIG. 6.14: Birefringence data of a FD-virus dispersion at a concentration of 230 c^*. Applied shear rate range from $\dot{\gamma} = 1\ s^{-1}$ and up to $\dot{\gamma} = 250\ s^{-1}$. Up to the third order of birefringence are visible. The alternating steps are due to shear in clockwise and counterclockwise.

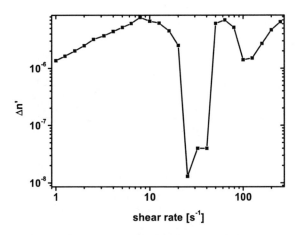

FIG. 6.15: Shear rate dependent birefringence extracted from Fig. 6.14.

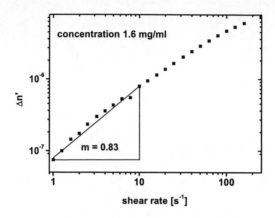

FIG. 6.16: Birefringence data of a FD-virus dispersion at a concentration of $40\ c^*$. Applied shear rate from $\dot{\gamma} = 1\ s^{-1}$ and up to $\dot{\gamma} = 250\ s^{-1}$. Detected birefringence maximum value was $\Delta n' = 7.5 * 10^{-8}$.

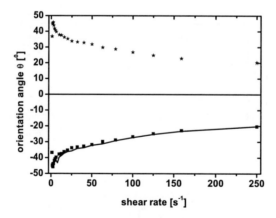

FIG. 6.17: The orientation angle data θ of a FD-virus dispersion at concentration of 40 c^*. The applied shear rates correspond to those used in Fig. 6.16. Note, stars and squares represent the two different angles from clockwise and counterclockwise measurements. The deviations in the values of the clockwise and counterclockwise measurements are below $\pm\ 2^\circ - 3^\circ$ for $\dot{\gamma} = 25\ s^{-1}$ (the star values are multiplied with -1).

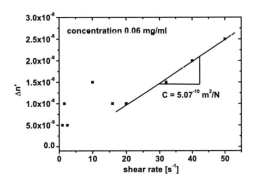

FIG. 6.18: The shear rate dependent birefringence $\Delta n'$ measurement of a FD-virus dispersion at a concentration of $1.5\ c^*$. The shear rates $\dot{\gamma}$ range from $1\ s^{-1}$ up to $250\ s^{-1}$. The smallest detected birefringence is around $\Delta'_{min} = 10^{-8}$. The stress optical coefficient for FD-virus was determine at this concentration to $C = 5.07 * 10^{-10}\ \frac{m^2}{N}$.

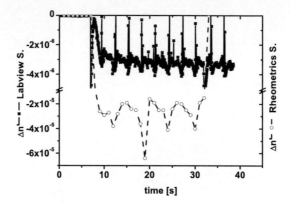

FIG. 6.19: Comparison of the birefringence $\Delta n'$ data measured with the commercial Rheometrics Software (lines with empty circles) and with the home written LabVIEW program (lines with filled squares). The differences are: 1st Overshoot during the start of shear is detectable, 2nd the factor of 10 times smaller signal, 3rd a faster data acquisition of the factor of 24, and 4th less fluctuations.

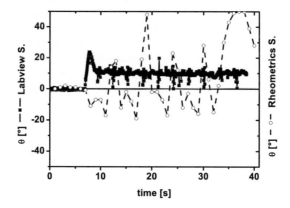

FIG. 6.20: Comparison of the orientation angle θ data measured with the commercial Rheometrics software (lines with open circles) and with the home written LabVIEW program (lines with filled squares). The differences are: 1st Overshoot during the start of shear is detectable, 2nd the factor of 10 times smaller signal, 3rd a faster data acquisition of the factor of 24, and 4th less fluctuations.

6.4 Conclusion of the chapter improvements in a rheo-optical set-up

The quality of the raw rheo-optical signal is strongly influenced by the background birefringence. The self-developed LabVIEW software, uses an on-the-fly FFT to determine the oscillation frequency of the spinner motor, circumvents problems of the commercial set-up and enables us to simultaneously acquire optical and mechanical data with higher a resolution. Using a variable retarder to compensate the background birefringence it is possible to reduce the negative effect of the background birefringence, by the factor of 20. It was also possible to increase the resolution in the time domain by the factor of 24. Furthermore it is possible to explain and simulate the effect of a background signal via a vectorial depiction. Analysing the signals of a rotating birefringent object then verified this effect and birefringence down to $\delta'_{min} = 10^{-8}$ could be detected. These experimental improvements were validated on two different set-ups measuring equivalent results. It should be possible to use this method on other commercial Optical Analysis Module hardware set-up's provided from Rheometrics or TA Instruments, if the here presented software would be incorporated.

Chapter 7

Mineralisation under shear

In the literature the influence of polymers on the mineralisation process of inorganic crystals is well described [Böhnlein-Ma 92, Liu 75, Öner 98]. Generally polymer is added in small amounts, with respect to the amount of the inorganic substance. During the mineralisation process the polymer influences the crystal growth. Polymers of different molecular weight and structure [Öner 98] are used to change the shape, size and aspect ratio of the growing crystallites. The idea within this chapter is to investigate the process of crystal growth under a defined shear field, with a focus on the resulting crystal habitus, e.g. size and aspect ratio.

7.1 Materials

The investigated mineralisation is the formation of zinc oxide (ZnO). The ZnO will crystallise homogeneously from an aqueous solution. Polycrystalline zinc oxide is widely used in vulcanisation processes, as UV-absorber, and fluorescent pigment in varistor ceramics, surface wave filters, gas sensors, and it doped with copper as catalyst for partial oxidations [Pearl 92, Solomon 93]. Its physical properties strongly depend on the grain size, on the dispersity, and on the contacts among grain boundaries of the crystallites. The polymers examined in this project are diblock-co-polymers of the structure polyethylene oxide-b-polymethacrylicacid (PEO-b-MA) see tab 7.1 and structure 7.1, and a statistical copolymer poly(ethyleneoxide-co-acrylicacid) P(EO-co-AA), see tab 7.2 and structure 7.2.

total molar mass M_w	PEO M_w	MA M_w	concentration
$3700 \frac{g}{mol}$	$3000 \frac{g}{mol}$	$700 \frac{g}{mol}$	$5 * 10^{-4} \frac{mg}{ml}, 2 * 10^{-3} \frac{mg}{ml}$

Table 7.1: Properties of polymer PEO-*b*-MA.

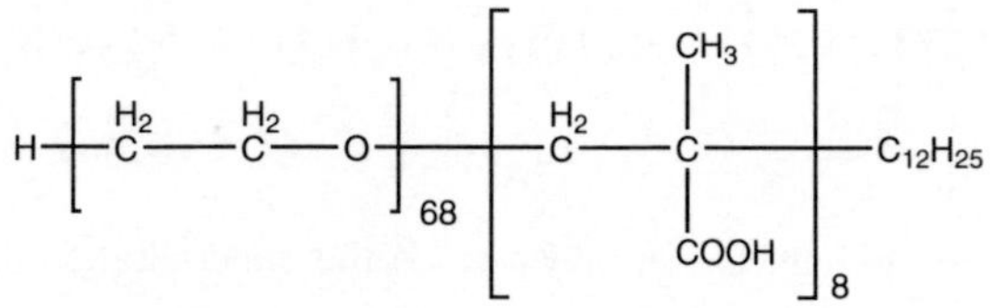

FIG. 7.1: Structure of the polymer PEO-b-MA.

total molar mass M_w	molar measure	concentration
$78800 \frac{g}{mol}$	$PEO : AA = 2 : 1$	$2 * 10^{-3} \frac{mg}{ml}$

Table 7.2: Properties of polymer P(EO-co-AA).

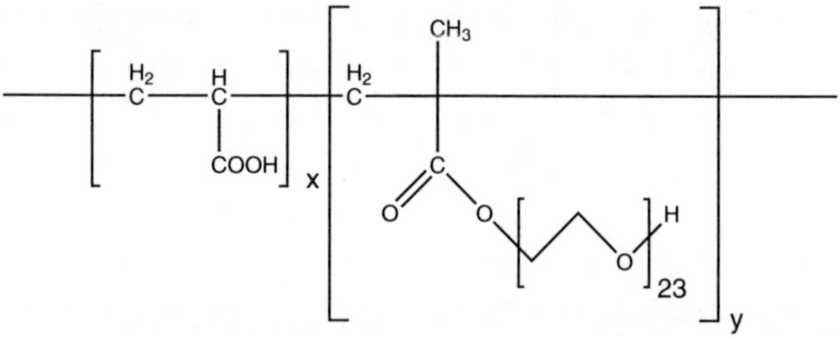

FIG. 7.2: Structure of the statistical copolymer P(EO-co-AA). The parameter x and y give the ratio of monomers within the polymer chain to be: $PEO : AA = 2 : 1$.

7.2 Experiments

The experiments are performed in a watery solution. The overall amount of the solution is about 5 ml. The Zinc nitrate ($Zn(NO_3)_2$), Urotropin (hexamethylenetetramine (HMT) $C_6H_{12}N_4$), and the polymer are mixed at room temperature to achieve a homogeneous mixture. While heating up the Urotropin disintegrates in formaldehyde H_2CO and ammonia NH_3.

$$(H_2C)_6N_4 + 6H_2O \longrightarrow 4NH_3 + 6H_2CO \qquad (7.1)$$

The ammonia is responsible for a homogeneous change of the pH-value simultaneously all over the solution. After the start of the applied shear the solution is heated for 60 minutes at 90 °C. No Zinc oxide is formed before the heating. Due to HMT the Zinc oxide is equally distributed over the whole sample (see Fig. 7.1). The concentrations used in the experiments are: Zinc nitrate 0.015 M, HMT 0.015 M, polymer PEO-b-MA $5 * 10^{-4} \frac{mg}{ml}$, $2 * 10^{-3} \frac{mg}{ml}$ and polymer P(EO-co-AA) $2 * 10^{-3}$ $\frac{mg}{ml}$. In these experiments a four times higher concentration of all ingredients is used and then compared with the normal experiments in the beaker. This is necessary due to the small sample volume (≈ 5 ml compared with a beaker of ≈ 300 ml) in the Couette that results in a very small amount of material after the experiment. These experiments are performed at different shear rates, using shear rates of 300 s^{-1}, 500 s^{-1}, 1000 s^{-1}, and 3000 s^{-1}. Additionally the reaction is also performed without shear and without any polymer and second without shear and with polymer as a reference. After initial tests with a water-bath and a solvent trap the set-up was changed because the evaporation could not be prevented for the 60 minutes time span needed for the experiment. Finally a Couette cell without solvent trap is used and water is continuously added with a syringe. In this way permanently a constant water level could be maintained. After the experiment the solution is extracted from the Couette cell with a syringe, to separate the solution in the gap from the one under the bob. This is done to check, if there are differences between these two solutions. In Fig. 7.3 one can see the area of the vertical gap where a defined shear field is applied (gap solution) and the area below the bob which is not sheared due to the air cushion between bob and cup. To reduce the amount of the non-sheared solution the vertical distance between bob and cup is minimised as much as possible to typically 100 μm. With this gap

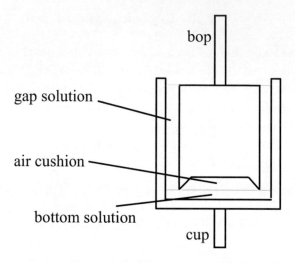

FIG. 7.3: Here a Couette Cell is shown, consisting of two parts, the upper bob and the lower cup. In the vertical gap the defined shear field is applied (gap solution). The non sheared solution is below the bob (bottom solution).

the volume of the unsheared solution can be neglected. The resulting solutions are first cooled with ice and then centrifuged. The crystallites are finally analysed via SEM.

7.3 Results and Discussion

The effect of shear on the crystallisation process is analysed by comparing the size, the shape and the aspect ratio of the particles. In the case of the low molecular weight polymer PEO-b-MA no influence is found see Fig. 7.4. Several concentrations combined with different shear rates are examined. The examined shear rates are: $100\ s^{-1}$, $300\ s^{-1}$, $500\ s^{-1}$, $1000\ s^{-1}$ and $3000\ s^{-1}$. The examined concentrations are: $5 * 10^{-4}\ \frac{mg}{ml}$ and $2 * 10^{-3}\ \frac{mg}{ml}$. At every applied shear rate about 30 crystallites were measured in length l, and diameter d. The average was taken to get representative numbers of the size of the crystallites. The principal effects of the addition of a polymer could only be seen in the change of the aspect ratio. Furthermore abrasive effects on the crystallites are found especially at higher shear rates. The results are presented in Fig. 7.4.

The influence of the high molecular weight polymer P(EO-co-AA) on the crystallisation process without any shear, results in a change of the size and of the aspect ratio. The length of the prolate like crystals is reduced by the factor of 1.9, whereas the diameter is not reduced. Therefore the aspect ratio is changed by the factor of 1.9 [Öner 98]. If one compares now the effect of shear on the crystallisation process the sample with polymer at no shear Fig. 7.5 (B) is the standard. Comparing graph (B) with (C) and (D) where shear rates of $300\ s^{-1}$ and $500\ s^{-1}$ are applied, a significant reduction by the factor 2 in both diameter and length is observed (see also table 7.3). If the shear rate is increased to values of $1000\ s^{-1}$ or $3000\ s^{-1}$ no change in size is found compared with the shear rates of $300\ s^{-1}$ and $500\ s^{-1}$. Only abrasive effects can be detected. The edges are less sharp and fewer twin crystals are found. One conclusion is that shear has an effect on the crystallisation process. In the examined case a reduction in size is detected. This is in contrast to experiments where no shear is applied and a change of the aspect ratio is found. This opens a pathway to easily modify crystals by the addition of polymer and / or by the application of shear. Changes in size and aspect ratio are the results. Due to the presented results, it does seem to be important to state that noticeable effects are only visible when the polymer has a high molecular weight. Therefore one can assume that the shear only has an effect on the polymer when the molecular weight is big enough. Under that condition the polymer chains, that are attached to the surface of the polymer, are long enough to be affected

shear rate	$0\ s^{-1}$
without polymer	$l = 1066\ nm,\ d = 200\ nm,\ a.r. = 5.3$
P(EO-co-AA)	$l = 567\ nm,\ d = 200\ nm,\ a.r. = 2.8$
shear rate	$300,\ 500\ s^{-1}$
without polymer	-
P(EO-co-AA)	$l = 293\ nm,\ d = 93\ nm,\ a.r. = 3.1$
shear rate	$1000,\ 3000\ s^{-1}$
without polymer	-
P(EO-co-AA)	like in $300,\ 500\ s^{-1}$

Table 7.3: Analysis of the length l, diameter d, and the aspect ratio $a.r.$ of the crystallites without polymer and with P(EO-co-AA)

by the applied shear field, and the crystallisation process is influenced. It can be assumed, that due to the shear the polymer which is attached to the crystallites will align itself parallel with the flow. This alignment is expected to influence the crystal growth and may therefore result in a different crystal habitus.

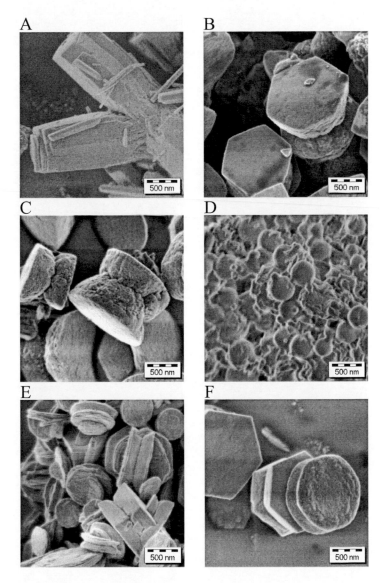

FIG. 7.4: SEM pictures of ZnO: (A) without shear and without polymer. (B)-(F) ZnO plus polymer PEO-b-MA at shear rate: $0\ s^{-1}$, $300\ s^{-1}$, $300\ s^{-1}$ (bottom solution), $1000\ s^{-1}$, $3000\ s^{-1}$. The total scale over the three black and two white bars are $500\ nm$.

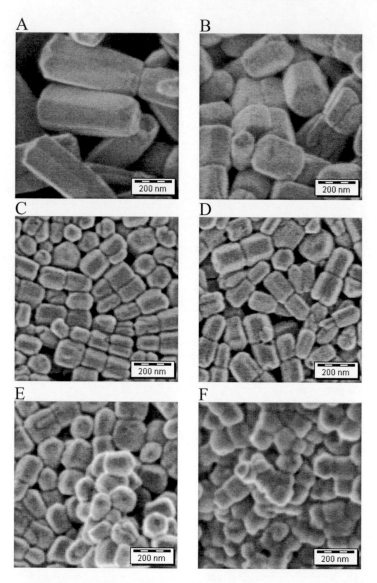

FIG. 7.5: SEM pictures of ZnO: (A) without shear and polymer. (B)-(F) ZnO including polymer P(EO-co-AA) at shear rate: $0\ s^{-1}$, $300\ s^{-1}$, $500\ s^{-1}$, $1000\ s^{-1}$, $3000\ s^{-1}$. The total scale over the three black and two white bars are $200\ nm$.

Chapter 8

Rheological behaviour of highly filled dispersions under steady shear and LAOS conditions

Complex fluids show a great variety of rheological properties [Macosko 94, Larson 99]. In this chapter the rheological analysis of dispersions, and specifically their behaviour under large oscillatory shear (LAOS) are presented and analysed. The rheological experiments cover the range from the linear to the non-linear regime. The conducted measurements are sweeps in the shear rate, sweeps in the frequency at constant strain amplitude and the sweeps in the strain at a constant frequency. Measurements of the viscosity, where a constant shear rate is applied, are steady state measurements, whereas the frequency and the strain sweeps are dynamic measurements. In the frequency sweep the strain amplitude is a parameter and the frequency the variable. In the strain sweep it is just the opposite, the frequency is the parameter and the strain amplitude the variable. Especially in the non-linear regime measurements at a constant frequency and a constant strain were used. Here frequency and strain amplitude are both parameters and are changed after each experiment to cover a broad range of frequencies combined with strain amplitudes. The standard analysis of the non-linear oscillatory data via FT-rheology is conducted in section 8.2.2, whereas a new analysis method of FT-rheological data is presented in chapter 9.

The samples examined in this chapter were synthesised via emulsion polymerisation. The instructions for the synthesis can be found in Appendix D. The

samples name contains the shortening CKA or CKB hinting on the lab journal A
or B, where the synthesis instructions were written down. The number gives the
consecutive numbers of experiments of synthesis performed in this work.

8.1   Linear rheology

The linear regime can be determined via a strain sweep, see Fig. 8.1, or a shear
rate sweep, see Fig. 8.2. A typical behaviour is the independence of the viscosity
(η, $|\eta^*|$) at small values of the shear rate or strain amplitude and then at higher
values the decrease of the viscosity with the increase of the shear rate or respectively with the strain amplitude. The cross-over from the Newtonian behaviour
to the shear thinning or strain softening behaviour is the begin of the non-linear
behaviour, or non-linear regime. This cross-over can be detected earlier by FT-
rheology than with the standard method, see Fig. 8.3. Characteristically in frequency sweeps dispersions show the change from an elastic response at small

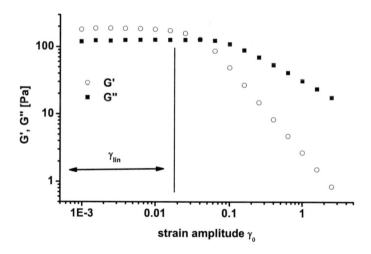

FIG. 8.1: G' and G" data of sample ckb117 determined via a strain sweep at a temperature
of 293 K: $\frac{\omega_1}{2\pi} = 1\ Hz$, $\gamma_0 = 0.0005 - 3$.

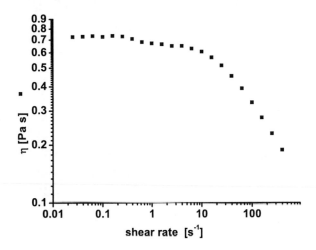

FIG. 8.2: Shear rate dependent viscosity η of the sample ckb117 at a temperature of 293 K: $\dot{\gamma} = 0.025\ s^{-1}$ to $400\ s^{-1}$.

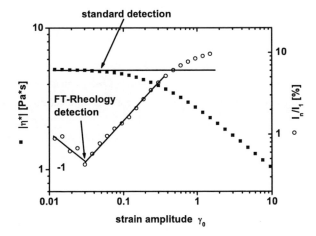

FIG. 8.3: η data of ckb103 determined via a shear rate sweep at a temperature of 293 K: $\dot{\gamma} = 0.025\ s^{-1}$ to $400\ s^{-1}$.

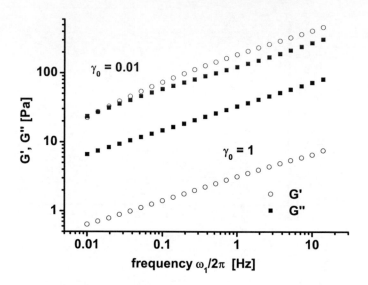

FIG. 8.4: G' and G'' data of ckb117 determined via a frequency sweep at a temperature of 293 K: 0.02 $Hz - 11\ Hz$, for $\gamma_0 = 0.01$ and $\gamma_0 = 1$.

strain amplitudes to an viscous response at larger strain amplitudes. Internal structures, like electrostatic repulsion or packing of the particles, result in a dominantly elastic response (see Fig. 8.4). Only after the application of a sufficient strain amplitude in dynamic experiments this elastic behaviour is overcome, resulting in a dominantly viscous response of the material. This was only observed in samples with a high solid content [Dames 01]. After dilution this effect is not observed anymore.

8.2 Non-linear rheology

In steady experiments dispersions typically show shear thinning behaviour. For the description of the shear thinning behaviour several descriptive models like Ostwald-de-Waele, Carreau or Cross are used. The Ostwald-de-Waele model equation (8.1) is used in the case where directly shear-thinning is observed, meaning no cross-over from the first Newtonian plateau can be detected:

$$\eta = b \cdot \dot{\gamma}^{-a}, \tag{8.1}$$

here a is the scaling parameter or shear thinning exponent, which can take values between 0 and 1. In the case of 0 the dispersion is Newtonian, whereas in the case of limiting value 1 the maximum shear thinning is found. As shown by equation (8.2) when $a = 1$ the force becomes independent of the shear rate and therefore constant:

$$\sigma = \eta(\dot{\gamma}) \cdot \dot{\gamma} = b \cdot \dot{\gamma}^{-1}\dot{\gamma} = b. \tag{8.2}$$

If additionally to the shear thinning the first Newtonian plateau is observed, the Carreau-model can describe the Newtonian and the shear thinning regime:

$$\eta(\dot{\gamma}) = \frac{\eta_0}{1 + (\beta\,|\dot{\gamma}|)^c}, \tag{8.3}$$

with η_0 (zero shear viscosity) the plateau value of the viscosity to the zero shear, c the scaling parameter with a value between 0 and 1, and with the inverse of β the pivot point (knee) of the curve. At this position of the pivot point the zero shear viscosity has dropped from the initial value in the Newtonian regime to $\frac{\eta_0}{2}$. Sometimes three parameter models can not describe the slope correctly. This is a problem of the fitting process, when at the cross-over from the first Newtonian plateau to the shear thinning regime a very sharp knee is found. Variations of the Carreau model with 4 parameters, like the Cross model are then used. The product of the exponents $c * d$, which is always smaller or equal to 1, describes the width of the knee, which means the transition from the slope 0 to the final slope:

$$\eta(\dot{\gamma}) = \frac{\eta_0}{[1 + (\beta\,|\dot{\gamma}|)^c]^d}. \tag{8.4}$$

These models work well for the samples examined within this thesis. Examples can be seen in the following chapters.

8.2.1 Non-linear behaviour under strain dependent LAOS conditions

Complex fluids, such as the dispersions examined within this work, show a great variety of rheological properties. Ahn et al. have conducted strain sweep tests on several different materials. They found four different classes of behaviour in strain sweep tests ([Hyun 02, Sim 03] visualised in Fig. 8.5. Type I is named strain thinning, when both G' and G'' show, after an initial Newtonian plateau a purely strain softening behaviour. Type II, called strain hardening, shows an increase in both moduli. Type III and type IV have both an overshoot in G'' at increasing strain amplitudes in common. Type III shows the overshoot in G'' only, whereas type IV shows a overshoot in G' too. Type I and type III were found in the experiments within this work. Strain softening behaviour, named type I, is believed to be caused by similar effects like the shear thinning. When applying a constant shear rate an alignment of microstructures with the flow reduces the

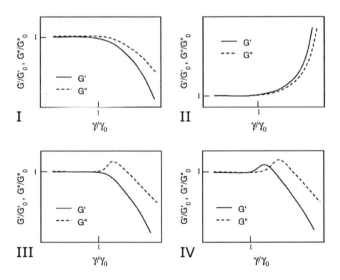

FIG. 8.5: The four different types of LAOS behaviour according to Ahn [Hyun 02], I) strain thinning, II) strain hardening, III) weak strain overshoot, and IV) strong strain overshoot.

local drag. The increasing alignment reduces the viscosity. The interparticular interactions are reduced due to the microstructural anisotropy coming from the applied strain. The particles in the dispersion interact in the Newtonian regime and rebuild their local structure faster than it is affected by the shear. With increasing applied strain the sum of the interactions are decreased and an alignment of the structures with the flow occurs [Hyun 02]. The moduli G' and G'' decrease further, the system becomes anisotropic. The type II behaviour was not found within this thesis, therefore it will not be treated here any further. The type III is characterised by a strain overshoot in G''. It is assumed to appear in systems with a weak network-like structure. Due to the electrostatic interactions and the higher mono dispersity of the particles in the highly concentrated dispersions a weak interaction is evoked.

During the application of an shear field this highly ordered structure resists its

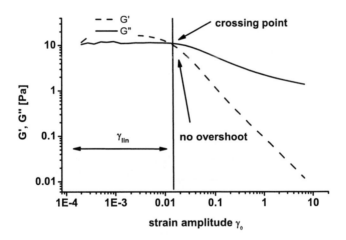

FIG. 8.6: Strain sweep plots from dispersion CKB171 (see Appendix D) synthesised via semi-continuous emulsion polymerisation, showing no overshoot in G' and therefore belonging to Ahn type I. The applied frequency was $\frac{\omega_1}{2\pi} = 1\ Hz$ at a temperature of $293\ K$. For $\gamma_0 < 2\%$ the sample responds linearly. At higher strain amplitudes the sample shows strain softening.

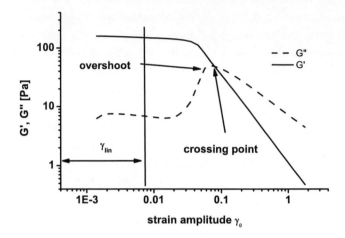

FIG. 8.7: Strain sweep plot at a frequency $\frac{\omega_1}{2\pi} = 1\ Hz$ and a temperature of $293\ K$ for dispersion CKB235, synthesised via mini-emulsion polymerisation. No overshoot in G' but in G'' is found and therefore this sample belongs to Ahn type III. For $\gamma_0 < 2\%$ the sample responds linearly. At higher strain amplitudes the sample shows strain softening.

deformation up to a certain strain. As a result the modulus G'' increases. This structure is destroyed by a large, critical deformation after which both moduli decrease. Note, as soon as there is a change in one of the moduli in samples showing Ahn type I or Ahn type III the loss tangents changes also. The type IV behaviour was not found either, and is therefore not treated here any further. In dispersions with high solid content, synthesised via semi-continuous emulsion polymerisation, see chapter 4.1, type I was exclusively found (see Fig. 8.6). In those samples no network-like structure is build up, due to the shorter Debye-length, that could resist the external flow field. In contrast to the dispersion synthesised via semi-continuous polymerisation, those synthesised via mini-emulsion polymerisation show the behaviour of type III. A weak network-like structure should be the reason according to Ahn. Here a typical example is given with the dispersion CKB 235 (see Fig. 8.7). In case of these dispersions, this could be explained by crystal-like structures established due to the higher mono dispersity of the particles. Additionally a far ranging electrostatic interaction potential, which is evoked by the longer Debye-length, adds to the network-like structure. These structures oppose the applied strain up to certain degree, after which they are overcome and destroyed.

8.2.2 Prediction and comparison of non-linearities under LAOS conditions

Analysing high solid content dispersions via FT-rheology was first introduced by Kallus et al. [Kallus 01]. In these studies on commercially available dispersions high non-linearities, respectively higher harmonics magnitudes, were found in samples with a high solid content. A measure for the non-linearity is the intensity of the third harmonic. In further experiments a decrease of the intensity of the third harmonic with an increasing salt content was found. Further examinations of polymer dispersions under LAOS conditions, performed by Carreau et al. [Craciun 03], confirmed the observation of decreasing intensities of the higher harmonics with increasing salt content. Additionally the strain dependence of the intensity on the higher harmonics showed a maximum. In the following the results from this analysis method will be shown. Due to the fact that this method considers the third and the fifth only, in some cases additionally the seventh and the ninth harmonic, a new analysis method was developed, see chapter 9. In the samples examined within the thesis many more higher harmonics were found. For compatibility with the analysis of others [Neidhöfer 03a], first the known analysis results will be shown, followed by the new method.

After the quantification of the non-linearities, a model is needed to predict the non-linear mechanical properties. As can be seen from oscillatory frequency data, G' and G" are dominantly viscous in the strain softening regime. In the Newtonian regime a dominantly viscous response could be found, in general at strain amplitudes larger than $\gamma_0 = 0.05$. Under the assumptions of an instantaneous adjustment to the applied shear and dominantly viscous response, a way to predict the non-linearities via the Carreau model should be possible. Starting from Newtons equation (8.5) to describe the viscous behaviour, the data of the viscosity η was emulated. The shear rate $\dot{\gamma}$ in Newtons equation, but also in the Carreau model (equation (8.3)), is given by equation (8.7):

$$\sigma = \eta(\dot{\gamma}) \cdot \dot{\gamma}, \tag{8.5}$$

$$\eta = \frac{\eta_0}{1 + (\beta \,|\dot{\gamma}|)^c}, \tag{8.6}$$

$$\dot{\gamma}(t) = \gamma_0 \omega_1 cos\omega_1 t. \tag{8.7}$$

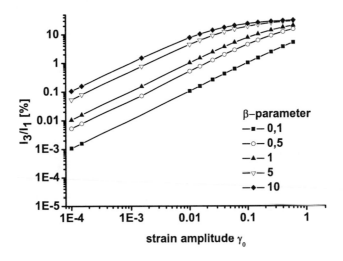

FIG. 8.8: Behaviour of the third harmonic magnitude depending on the strain amplitude γ_0. The parameters β ranging from $0.1 - 10$ and the parameter c, here set to 1, are from equation (8.6) and the excitation frequency $\frac{\omega_1}{2\pi} = 1\,Hz$ is taken from equation (8.7).

The parameters η_0, β and c needed for the Carreau model were extracted from the steady shear viscosity. The emulated data is Fourier transformed and the results are compared with the measured non-linearities. For a set of parameters these calculations have been done to give a broad set of simulations. Based on these simulations an evaluation of the non-linearities depending on η, β or c is possible. Note, the shape and the relative intensities of the higher harmonics do not depend on the zero shear viscosity η_0. The behaviour of the predicted relative third and fifth harmonic in amplitude depending from β is shown in the plots Fig. 8.8, and Fig. 8.9, whereas the dependence on the parameter c is shown in Fig. 8.10, and Fig. 8.11. It can be stated that for all sets of parameters a strictly monotonic increasing function for increasing strain amplitudes is found. A higher β results in higher intensities of the harmonics, at a constant value for c, until a maximum is reached. A value of 33 %, which is the maximum for the intensity of the 3rd-harmonic, is already reached at strain amplitudes of about 0.3 for a

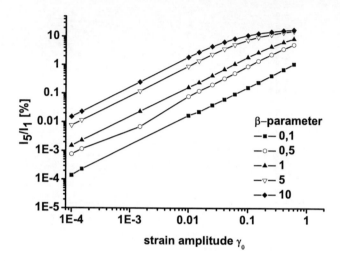

FIG. 8.9: Behaviour of the fifth harmonic magnitude depending on the strain amplitude γ_0. The parameters β ranging from $0.1 - 10$ and the parameter c, here set to 1, are from equation (8.6) and the excitation frequency $\frac{\omega_1}{2\pi} = 1\,Hz$ is taken from equation (8.7).

value of $\beta = 10s$. The intensities of the 3rd harmonic are always larger than the intensities of the 5th-harmonic, with a maximum value of 20 %. For a constant value of β the variation of c results in a change of the slope of the intensities of the higher harmonics. With an increasing value for c an increase of the slope is found. For all values of c the value of the intensities of the 3rd-harmonic are larger than the intensities of the 5th-harmonic. The relative behaviour of the phases of the odd harmonics can be extracted from the shear stress σ based on Newton. In Newtons equation, the viscosity is described by Carreau, and the shear rate is given by $\dot{\gamma} = \gamma_0\omega_1 cos\,(\omega_1 t)$.

$$\sigma = \eta \times \dot{\gamma} = \frac{\eta_0\gamma_0\omega_1 cos\,(\omega_1 t)}{(\beta\,|\gamma_0\omega_1 cos\,(\omega_1 t)|)^c}.\qquad(8.8)$$

This results, with the parameter $c = 1$ to:

$$\sigma = \frac{\eta_0}{\beta}\frac{cos\,(\omega_1 t)}{|cos\,(\omega_1 t)|}.\qquad(8.9)$$

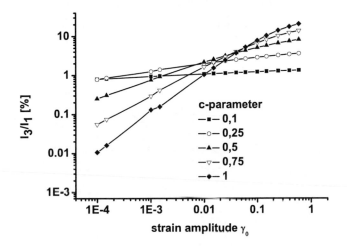

FIG. 8.10: Behaviour of the third harmonic magnitude depending on the strain amplitude γ_0. The parameters c ranging from $0.1 - 1$ and the parameter β, here set to $1\ s$, are from equation (8.6) and the excitation frequency $\frac{\omega_1}{2\pi} = 1\ Hz$ is taken from equation (8.7).

The fraction of the cosine and the absolute cosine function can be described via the rectangle function, expanded here as a Fourier series:

$$\frac{cos\,(\omega_1 t)}{|cos\,(\omega_1 t)|} = \frac{4}{\pi}\left(sin\,[\omega_1 t] + \frac{1}{3}sin\,[3(\omega_1 t)] + \frac{1}{5}sin\,[5(\omega_1 t)] + ...\right), \qquad (8.10)$$

$$\sigma = \frac{\eta_0}{\beta}\frac{4}{\pi}\left(sin\,[\omega_1 t] + \frac{1}{3}sin\,[3(\omega_1 t)] + \frac{1}{5}sin\,[5(\omega_1 t)] + ...\right). \qquad (8.11)$$

Under the assumption of a purely viscous response a shift in the time of 90 °, corresponding to $\frac{\pi}{2}$, is applied:

$$\omega_1 t -> \omega_1 t + \frac{\pi}{2}, \qquad (8.12)$$

$$\sigma = \frac{\eta_0}{\beta}\frac{4}{\pi}\left(sin\,\left[\omega_1 t + \frac{\pi}{2}\right] + \frac{1}{3}sin\,\left[3(\omega_1 t + \frac{\pi}{2})\right] + \frac{1}{5}sin\,\left[5(\omega_1 t + \frac{\pi}{2})\right] + ...\right), \qquad (8.13)$$

where z is:

$$\sigma = \frac{\eta_0}{\beta}\frac{4}{\pi}(z), \qquad (8.14)$$

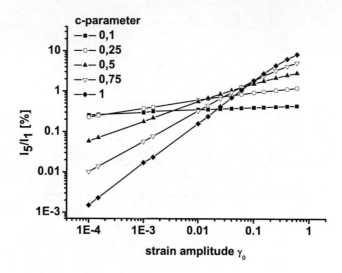

FIG. 8.11: Behaviour of the magnitude of the fifth harmonic depending on the strain amplitude γ_0. The parameters c ranging from $0.1 - 1$ and the parameter β, here set to $1\ s$, are from equation (8.6) and the excitation frequency $\frac{\omega_1}{2\pi} = 1\ Hz$ is taken from equation (8.7).

$$z \propto sin\left[\omega_1 t + \frac{\pi}{2}\right] + \frac{1}{3}sin\left[3(\omega_1 t + \frac{\pi}{2})\right] + \frac{1}{5}sin\left[5(\omega_1 t + \frac{\pi}{2})\right] + ... \qquad (8.15)$$

$$z \propto cos\left[\omega_1 t\right] + \frac{1}{3}sin\left[3\omega_1 t + \frac{3}{2}\pi\right] + \frac{1}{5}sin\left[5\omega_1 t + \frac{5}{2}\pi\right] + ... \qquad (8.16)$$

$$z \propto cos\left[\omega_1 t\right] + \frac{1}{3}sin\left[3\omega_1 t + \frac{1}{2}\pi + \pi\right] + \frac{1}{5}sin\left[5\omega_1 t + \frac{1}{2}\pi + 2\pi\right] + ... \qquad (8.17)$$

$$z \propto cos\left[\omega_1 t\right] + \frac{1}{3}sin\left[(3\omega_1 t + \pi) + \frac{1}{2}\pi\right] + \frac{1}{5}sin\left[(5\omega_1 t + 2\pi) + \frac{1}{2}\pi\right] + ... \qquad (8.18)$$

$$z \propto cos\left[\omega_1 t\right] + \frac{1}{3}cos\left[3\omega_1 t + \pi\right] + \frac{1}{5}cos\left[5\omega_1 t + 2\pi\right] + ... \qquad (8.19)$$

In the final equation (8.20) the relative phase shift between the fundamental and

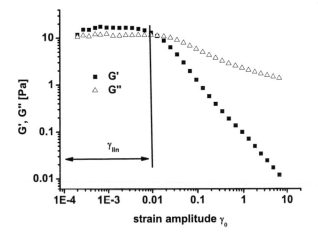

FIG. 8.12: Strain sweep of sample CKB171 for the determination of G' and G" at a frequency of $\frac{\omega_1}{2\pi} = 1\ Hz$, and a temperature of $293\ K$.

the phases of the higher harmonics are given by:

$$\sigma \propto \frac{\eta_0}{\beta} \frac{4}{\pi} \left(cos \left[\omega_1 t + \underbrace{0\pi}_{0°} \right] + \frac{1}{3} cos \left[3\omega_1 t + \underbrace{\pi}_{180°} \right] + \frac{1}{5} cos \left[5\omega_1 t + \underbrace{2\pi}_{360°} \right] + ... \right)$$

(8.20)

For the third harmonic the calculated relative phase results in $180°$. For the fifth harmonic the phase is given by $360°$, corresponding to $0°$. For the seventh harmonic a value of $540°$ is found, that corresponds to $180°$. The maximum values of the higher harmonics intensities were not reached in the experiments, because the simplification of using $c = 1$ is not always valid and because a possible elastic contribution is not considered. These facts should be considered while interpretation of the behaviour of the dispersions, e.g. the value of c could be in the range of $0.8 - 1$. The non-linear regime, and the dominantly viscous response is determined via a strain sweep (see Fig. 8.12). At strain amplitudes larger than $\gamma_0 = 0.1$ the behaviour is dominantly viscous. Afterwards the FT-rheology was applied. The time domain and magnitude spectra of the sample ckb171 are representative for the results acquired via the FT-rheology (see Fig. 8.13). First

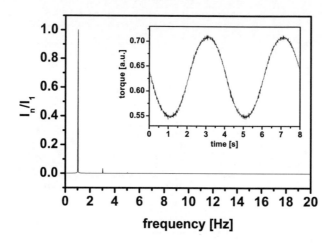

FIG. 8.13: Time domain data (inset) and frequency domain data of sample CKB171 at a frequency of $\frac{\omega_1}{2\pi} = 1\ Hz$, a strain amplitude $\gamma_0 = 1.6$, and a temperature of 293 K.

the results from the dispersions synthesised via semi-continuous emulsion polymerisation are analysed. Typical plots of the magnitude and phase spectra are presented. The predictions of the magnitudes, either created in experiments at a constant frequency and changing strain amplitude or the other way round, are in good quantitative and qualitative agreement as can be seen in Fig. 8.14 and Fig. 8.15. The phases and their predictions overlay very well, resulting in 0 ° for the third and seventh, and 180 ° for the fifth and the ninth harmonic (see Fig. 8.16 and Fig. 8.17). The phase behaviour, with a value of 180 ° for Φ_3, is, according to the phase analysis of Neidhöfer [Neidhöfer 03a], defined as strain softening. For the increase of the higher harmonic phases the value for the strain softening behaviour changes by 180 ° giving for the fifth harmonic 0 ° and for the seventh again 180 °. As mentioned above exactly these results were found. With the above shown results it can be stated that the approach for the emulation of the data is applicable. The assumption, of an instantaneous adjustment, and that the sample shows no memory, is within the investigated dispersions correct.

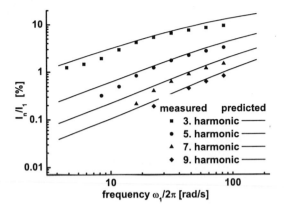

FIG. 8.14: Comparison of the prediction of higher harmonics, extracted via the Carreau model (see equation (8.6)), with the measured higher harmonics magnitudes, acquired via the FT-rheology. The frequency dependence of the sample CKB171, measured at a strain amplitude of $\gamma_0 = 0.1$, and a temperature of 293 K. The values from the Carreau fit are for $\eta_0 = 3.11\ Pas$, for $\beta = 9.23\ s$, and for $c = 0.58$.

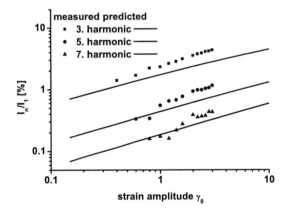

FIG. 8.15: Comparison of the prediction of higher harmonics, extracted via the Carreau model (see equation (8.6)), with the measured higher harmonics magnitudes, acquired via the FT-rheology. The strain dependence of the sample CKB171, measured at a frequency of $\frac{\omega_1}{2\pi} = 1\ Hz$, and a temperature of 293 K. The values from the Carreau fit are for $\eta_0 = 3.11\ Pas$, for $\beta = 9.23\ s$, and for $c = 0.58$.

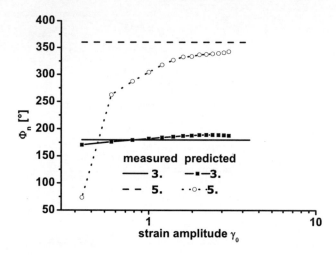

FIG. 8.16: Predicted and measured values for the phases of the 3rd and 5th harmonic. Predictions overlay well, especially at higher strain amplitudes. The sample CKB171 was measured at a frequency of $\frac{\omega_1}{2\pi} = 1\ Hz$, and a temperature of $293\ K$.

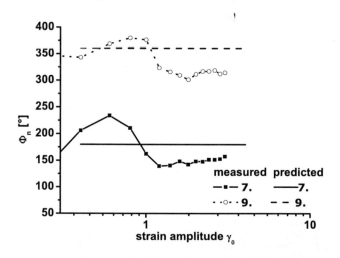

FIG. 8.17: Predicted and measured values for the phases of the 7rd and 9th harmonic. Predictions overlay well, especially at higher strain amplitudes. The sample CKB171 was measured at a frequency of $\frac{\omega_1}{2\pi} = 1\ Hz$, and a temperature of $293\ K$.

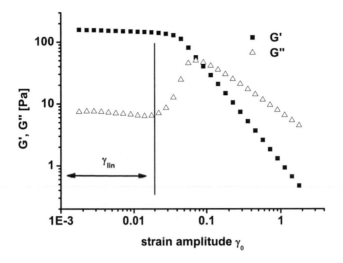

FIG. 8.18: Strain sweep of sample CKB235 for the determination of the linear regime and of G' and G" at a frequency of $\frac{\omega}{2\pi} = 1\ Hz$ was applied at a temperature of 293 K.

The non-linear regime of dispersions, synthesised via mini-emulsion polymeri-sation, were determined via a strain sweep see Fig. 8.18. The typical strain am-plitude dependence of magnitudes of the higher harmonics in dispersions synthe-sised via mini-emulsion polymerisation, see Fig. 8.19. The measured magnitude and the prediction of the higher harmonics magnitudes overlay well, like in the case of the dispersions synthesised differently (see Fig. 8.15). Differences were found in the phase Φ_n (see Fig. 8.20). At large strain amplitudes the values of the phases are above 220 ° for the 3rd phase, and above 50 ° for the 5th phase, where phase values of 180 ° and 0 ° were expected. The complex viscosity $|\ \eta^*\ |$ shows at the same deformation strain softening behaviour, suggesting a phase Φ_n of 180 ° or 0 °. These samples show a different phase behaviour, than is expected by the description of Neidhöfer, where for a strain softening behaviour a phase value of 180 ° for the 3rd harmonic is predicted. The offset of approximately 40 ° − 50 ° for Φ_3 and Φ_5 is not properly understood. A possible explanation could be,

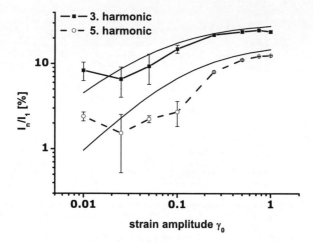

FIG. 8.19: The strain dependence of the magnitudes of the dispersion CKB235, which is typical for dispersions synthesised via mini-emulsion polymerisation. The magnitudes of the 3rd and the 5th harmonics are shown here. The error bars indicate the reproducibility after 3 measurements. The lines show the predicted behaviour of the higher harmonics. A frequency of $\frac{\omega_1}{2\pi} = 1\ Hz$ was applied at a temperature of $293\ K$. The values from the Carreau fit are for $\eta_0 = 3000\ Pas$, for $\beta = 200\ s$, and for $c = 0.94$.

that the assumption of a purely viscous response is not valid, meaning also elastic contributions are found. Another reason could be that the parameter c is assumed to be always 1. Additionally to these deviations in the values of the phases Φ_n a 2nd harmonic is found (see Fig. 8.21 for the magnitude plot and see 8.22 for the phase plot) that cannot be neglected since the values of the magnitude are larger than 3 %. It is not clear, what influence the appearance of the 2nd harmonic has on the 3rd and 5th phase to give an offset of 50 °. Furthermore the reason for the appearance of the 2nd harmonic could only be found in the different method of synthesising, which results in different properties of the dispersions e.g. the higher mono dispersity, the longer Debye-length (The Debye-length is in the range of $1.7\ nm - 1.8\ nm$ for dispersions synthesised via mini-emulsion-polymerisation compared with $0.3\ nm - 0.4\ nm$ for the dispersions for those synthesised via

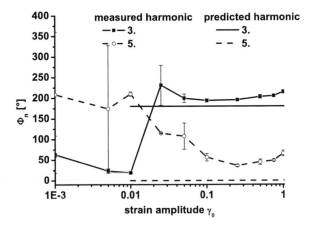

FIG. 8.20: The phases of the 3rd and the 5th harmonic of the dispersion CKB235 under application of the same experimental conditions and simultaneously detected like in Fig. 8.19.

semi-continuous emulsion-polymerisation) and the higher zero shear viscosity η_0 at the same volume fraction. The effective radius of the particle is larger due to the longer Debye-length $1.7\ nm - 1.8\ nm$. Under quiescent conditions the particles will pack, due to the high mono dispersity, in a more regular way. When a strain is applied the bulk set of particles will break up. At this point one could imagine a break up into layers, with the solvent flowing in between. The larger hydrodynamic radius, originating from the longer Debye-length, could play an important role. The potential is more far ranging than in the dispersions synthesised otherwise, resulting in a higher viscosity. The particles can break up into layers by the applied shear and leave an intermediate free place for the solvent. This solvent layer will then reduce the viscosity. An explanation for the appearance of the 2nd harmonic could be shear bands. In simulations and measurements [Graham 95, Heymann 01, Keunings 04, Sagis 01, Sim 03] second harmonics occurred, in e.g. polymer solutions, due to wall slip. The second harmonic can then result in a response signal such as a saw tooth. Simulations still have to prove

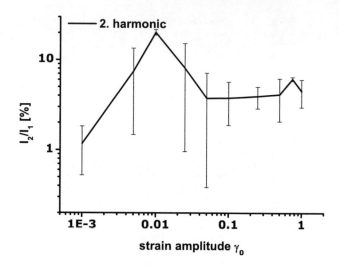

FIG. 8.21: The strain dependence of the magnitudes of the dispersion CKB235 synthesised via mini-emulsion polymerisation. Here the 2nd harmonic is shown. A frequency of $\frac{\omega_1}{2\pi} = 1\ Hz$ was applied at a temperature of $293\ K$.

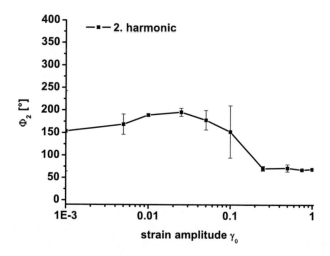

FIG. 8.22: Corresponding 2nd phase to Fig. 8.21.

that shear bands in dispersions show even higher harmonics.

8.3   Conclusion of the chapter rheological behaviour of highly filled dispersions under steady and LAOS conditions

The analysis of the rheological behaviour of polymer dispersions under linear and non-linear shear conditions, with an emphasis on FT-rheology on model systems, was conducted. Two different kinds of polystyrene dispersions, stabilised with acrylic acid, were examined. The differences result from two synthesis methods of the emulsion polymerisation. One method performs this radical reaction in one step, whereas the other method uses two steps, where in the second step the reactive ingredients are permanently added. With both methods dispersions with a very high solid content ($\varphi < 0.3$) were synthesised. The particle size is within the same range of $85\ nm - 170\ nm$. Deviations in the polydispersity (0.3 for semi-continuous, and 0.01 mini-emulsion polymerisation) and the Debye length (0.3 nm for semi-continuous synthesis and $1.8\ nm$ for mini-emulsion polymerisation) are the major differences. In the non-linear regime the samples are analysed via LAOS experiments. In the strain dependence of G' and G'', an overshoot in G'' could be observed in the samples synthesised via mini-emulsion polymerisation. This could be explained by the larger Debye-length, that leads to far ranging forces, which induce an order in these systems. Therefore dispersions, synthesised via semi-continuous emulsion polymerisation, with a shorter Debye-length, do not show this overshoot.

Differences between the two model systems could also be found analysing the mechanical higher harmonics (FT-rheology). In both model systems higher harmonics could be detected. Assuming an instantaneous adjustment and a dominantly viscous response, these higher harmonics could be predicted via the method explained in detail in chapter 8.2.2. In the dispersions synthesised via mini-emulsion polymerisation additionally a second harmonic was detected. Furthermore the phases of the third and fifth harmonic were not at expected values of $180\ °$ and $0\ °$, which are assigned to strain softening behaviour, but a deviation of about $20\ ° - 50\ °$ is experimentally detected. In recent literature [Graham 95, Heymann 01, Keunings 04, Sagis 01, Sim 03] the appearance of even harmonics is either detected or explained by shear bands. Wall slip is additionally known to produce second harmonics.

Chapter 9

Separation of LAOS-response into characteristic response functions

The analysis of non-linear oscillatory mechanical data with respect to the amplitudes and the phases of the higher harmonics, does not always result in a simple physical interpretation. In the standard analysis of non-linear oscillatory data used so far, the focus is on the phase and the magnitude of the third harmonic. In many cases this analysis method is justified due to the fact that only the intensity of the third harmonic is large enough to be analysed. On the other side in samples like the polystyrene dispersions examined here, a large number of higher harmonics of the excitation frequency with high intense harmonics are detected. In such samples not only odd harmonics are observed, but also even harmonics. These even harmonics are not negligible because their intensities are larger than 1 %, see Fig. 9.1. If one focusses on the third harmonic only, a lot of information is not considered in these polystyrene dispersions. Therefore a method is required that considers the whole overtone spectra. This is done by considering the spectrum as a superposition of different overtone spectra of typical non-linear rheological effects, like strain hardening, strain softening and shear bands or wall slip. The functions used to describe known rheological phenomena will be called characteristic functions, see chapter 9.1.

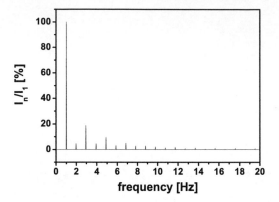

FIG. 9.1: FT magnitude spectra of the sample ckb222 at an excitation frequency of $\frac{\omega_1}{2\pi} = 1\ Hz$ and a temperature of $293\ K$.

9.1   Introduction of an analysis method for LAOS-response based on characteristic response functions

The quantification of the non-linear response with respect to these characteristic functions, e.g. strain hardening, strain softening under LAOS via FT-rheology, is based on the determination of the magnitude of the higher harmonics ($\frac{I_n}{I_1}$), relative to the fundamental and the determination of the corresponding phases Φ_n.

The measured signal will be split up into four fundamental contributions, see equation (9.1) to equation (9.4) and Fig. 9.2. These contributions are a sinusoidal functions, see equation (9.1), describing the linear contribution. A rectangle function, see equation (9.2), describing the strain softening contribution. A triangular function, see equation (9.3), describing the strain hardening. Finally a sawtooth function, see equation (9.4), describing shear bands or wall slip. These functions can each be varied via both the amplitude and the time-lag respective to the others. The time domain data and the magnitude spectra of these character-istic functions: the linear response, the rectangular wave, the triangular, and the sawtooth wave are shown in Fig. 9.2. In the following the Fourier-series of these characteristic functions are introduced.

The linear response is given by:

$$\sigma_l(t) = A_l sin(\omega t + \delta_l) \tag{9.1}$$

The periodic rectangle function is given by:

$$\sigma_r(t) = A_s \frac{4}{\pi} \left(sin(\omega t + \delta_r) + \frac{sin3(\omega t + \delta_r)}{3} + \frac{sin5(\omega t + \delta_r)}{5} + ... \right) \tag{9.2}$$

The periodic triangle is given by:

$$\sigma_t(t) = A_t \frac{4}{\pi} \left(sin(\omega t + \delta_t) - \frac{sin3(\omega t + \delta_t)}{3^2} + \frac{sin5(\omega t + \delta_t)}{5^2} - ... \right) \tag{9.3}$$

The periodic sawtooth is given by:

$$\sigma_{st}(t) = A_{st} 2 \left(sin(\omega t + \delta_{st}) - \frac{sin2(\omega t + \delta_{st})}{2} + \frac{sin3(\omega t + \delta_{st})}{3} - ... \right) \tag{9.4}$$

$$\sigma(t) = \sigma_l(t) + \sigma_r(t) + \sigma_t(t) + \sigma_{st}(t) \tag{9.5}$$

By superimposing these different contributions (equation (9.5)) the measured time domain signal is then reconstructed. Additionally the measured oscillatory signal is analysed with respect to the higher harmonics intensities I_n and phases Φ_n. Both the reconstructed time data and the FT analysis of the experimental and reconstructed signal are used to determine the different contributions of strain softening and hardening. In case where even harmonics are found a sawtooth function is used to describe the even harmonics. These even harmonics might be caused by shear bands or wall slip. The superposition of these four functions results in a time response that should mimic the measured time domain signal. The different amplitudes and phases of the functions are manually optimised. Since the time signal is often twisted, the reconstructed signal has to be adapted by changing the relative phases of the overlaying functions (sine, sawtooth...) with respect to each other. After the Fourier transformation the magnitudes and the phases of the simulated and of the measured signal should be identical and the time response should perfectly overlay.

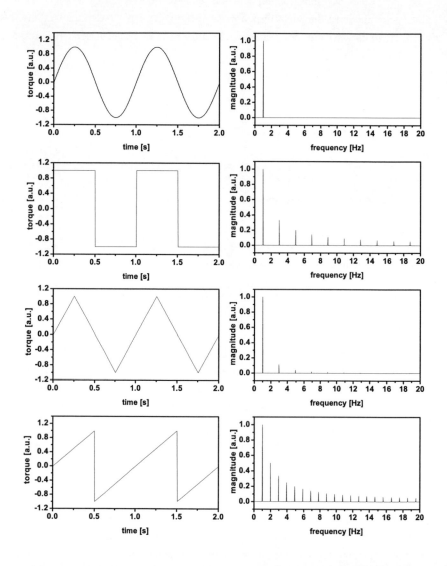

FIG. 9.2: Plots showing the four characteristic functions, the time domain signals of a sine, a rectangular, a triangular and a sawtooth shaped wave, as the characteristic response function of linear response, strain softening, strain hardening, and shear bands on the left and their corresponding FT magnitude spectra on the right.

phase lag	I_3/I_1 [%]	φ_3 [°]
0 °	33.3	180
110 °	51	82
140 °	63	116

Table 9.1: Superposition of a simulated rectangle function and a simulated sawtooth function. The resulting intensities and phases of the third harmonic, depending on the phase lag between the rectangle function and the sawtooth function are presented. The amplitude is 1 for both the rectangle and the sawtooth function.

For a better understanding the Fourier spectra of the characteristic functions are now introduced. The linear contribution, a sine wave, shows in the Fourier spectra one peak at the excitation frequency only. The strain softening contribution, a rectangle function, and the strain hardening wave, a triangle function, show only peaks at **odd** higher **harmonics** of the excitation frequency. Their main difference is the decrease of the intensity of the higher harmonics with increasing frequencies. The intensities of the higher harmonics decrease in the case of strain softening with $\frac{1}{n}$ ($n = 1, 3, 5,...$), and in the case of strain hardening with $\frac{1}{n^2}$ ($n = 1, 3, 5,...$). The last characteristic function introduced here is used to mimic the shear band or wall slip contribution, a sawtooth wave. This contribution has higher harmonics at **even** and **odd** multiples of the excitation frequency. The intensity decreases with a factor of $\frac{1}{n}$ ($n = 1, 2, 3,...$). The interaction of the different contributions can result in an increase or a decrease of the intensity of harmonics depending on the phase lag between the different contributions. This phase lag between the contributions can also affect the phase of the higher harmonics. As an example the behaviour of the phase and the magnitude of the third harmonics originating from a superposition of a rectangle function and a sawtooth function is simulated. The values for the intensity of the third harmonic and phase of the third harmonic are listed depending on the phase lag between the rectangle function and the sawtooth wave, see tab 9.1. It can clearly be seen that the phase lag can have a pronounced influence on the values of the intensities and phases.

To illustrate this method the samples ckb222 and ckb229 have been analysed according to this separation method. The time domain signal and the magnitude spectra of these two different samples are plotted in Fig. 9.3. The different con-

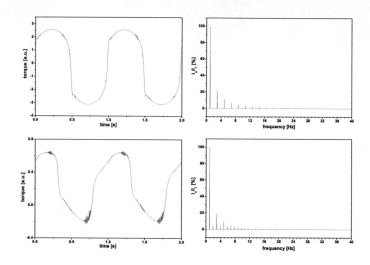

FIG. 9.3: The time domain signals of the samples ckb222 and ckb229 on the left, and the FT magnitude spectra are shown on the right. The time domain signals were recorded at $298\ K$ with $\frac{\omega_1}{2\pi} = 1\ Hz$ and for ckb222 $\gamma_0 = 1$ and for ckb229 $\gamma_0 = 6$.

tributions and the resulting data from the sample ckb222 according to tab 9.3 is presented in Fig. 9.4. In Fig. 9.5 and Fig. 9.6 the overlay of the measured data and the reconstructed data is presented. As can be seen the curves overlay well in the maxima but the deviations between measured and reconstructed data increases around the turning point. In tables 9.3 and 9.2 the different contributions of the reconstruction, their amplitudes, their phase, and the time-lag, are presented. The time-lag is derived from the phase shift between the linear contribution and the strain softening or shear band contribution.

For a better comparison the frequency domain data of the measured and reconstructed higher harmonics magnitudes and phases of the samples ckb222 and ckb229 are given in table 9.4 and table 9.5. Good agreement, typically within less than 1% relative deviation can be found for $\frac{I_2}{I_1}$, $\frac{I_3}{I_1}$, and $\frac{I_5}{I_1}$ in the magnitude spectra. For $\frac{I_2}{I_1}$, $\frac{I_3}{I_1}$, and $\frac{I_5}{I_1}$ the values for the phase give also very reasonable results, generally within less than 10 °. The higher order harmonics are in a less

	frequency $[Hz]$	amplitude $[a.u.]$	phase $[°]$	time-lag $[ms]$
sinusoidal	1	1	0	0
rectangle	1	1.47	35	97
sawtooth	1	0.0025	76	211

Table 9.2: Frequency, intensity and phase of the reconstructed response of sample ckb229, see Fig. 9.6.

	frequency $[Hz]$	amplitude $[a.u.]$	phase $[°]$	time-lag $[ms]$
sinusoidal	1	1	0	0
rectangle	1	0.85	43	119
sawtooth	1	0.2	22	61

Table 9.3: Frequency, intensity and phase of the reconstructed response of sample ckb222, see Fig. 9.5.

good agreement for the intensities. One reason for this effect is the rectangular wave used for the simplification. This function has a very steep slope (infinity) that results in very sharp flanks in the resulting reconstructed signal. Here a trapezoidal function could help to reduce the deviations. Similarly the sharp cut-off in the saw-tooth function should be smoothed. An other problem is that the intensities of the reconstruction result generally in higher intensities than the measured ones. This is a consequence of the simplification (sinusoidal, rectangular, triangular, and sawtooth shape) and the related maximum possible non-linearities. Obviously these maximum possible non-linearities are rarely reached within real materials. Simple multiplication with e.g. a Gaussian function in the reconstructed FT-spectra might therefore be used to get a better agreement for the higher harmonics intensities.

Irrespective of the minor problems related to the higher order intensities the phase of a third harmonic not being either 0 ° or 180 ° can now be seen as a superposition of different contributions. This phase can then be analysed by a superposition of a third harmonic originating from a strain softening process and a shear banding process in the sample. These different processes have furthermore a phase related to the time-lag of these different contributions, see tab 9.3, and tab 9.2.

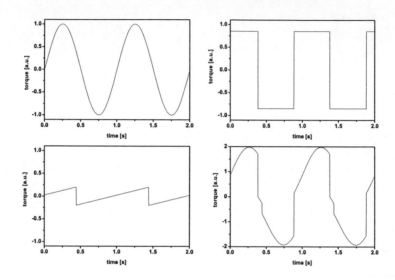

FIG. 9.4: In these plots the three different contributions and their sum for the reconstruction of the time domain signal of the sample ckb222 are shown. The sine-function has an
amplitude of 1 and a phase angle 0 °, the rectangle function has an amplitude of 0.85 and
a phase angle of 43 °, and the sawtooth has an amplitude of 0.2 and a phase of 22 °, see
tab 9.3.

harmonics	2.	3.	4.	5.	6.	7.
measured magnitude [%]	4.7	19.4	4.7	9.7	3.2	4.9
reconstructed magnitude [%]	4.0	19.4	2.0	12.3	1.3	10.3
measured phase [°]	214	236	76	92	305	305
reconstructed phase [°]	213	236	87	108	101	325

Table 9.4: The values for the magnitudes and the phases from the measurements and the
reconstruction of the sample ckb222.

In a following rectangle this analysis was then applied to a whole data set of
oscillatory response signals. These data sets originate from non-linear measure-

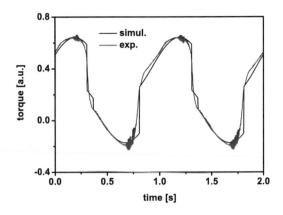

FIG. 9.5: Overlay of the torque response data and the corresponding reconstruction for
the sample ckb222.

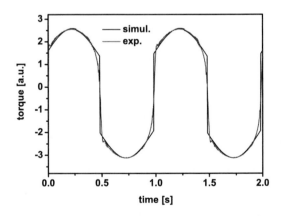

FIG. 9.6: Overlay of the torque response data and the corresponding reconstruction for
the sample ckb229.

harmonics	2.	3.	4.	5.	6.	7.
measured magnitude [%]	0.2	21.8	0.1	11.3	0.1	7.2
reconstructed magnitude [%]	0.0	21.7	0.0	13.0	0.0	9.3
measured phase [°]	64	177	266	351	141	163
reconstructed phase [°]	76	177	27	356	41	175

Table 9.5: The values for the magnitudes and the phases from the measurements and the reconstruction of the sample ckb229.

ments at increasing strain amplitudes. In a strain sweep these samples show the Ahn III behaviour, see Fig. 9.7 and Fig. 9.8. In Fig. 9.9 and Fig. 9.11 the strain amplitude dependence of the amplitudes of the samples ckb222 and ckb229 are plotted, whereas in Fig. 9.10 and Fig. 9.12 the strain amplitude dependence of the phases are presented. In tab 9.6, and tab 9.7 the values for the amplitudes and phases of the rectangle function and the sawtooth function of the samples ckb229 and ckb222 are given. For both samples the characteristic functions for the linear response, strain softening, and wall slip or shear bands are found. The triangle function, describing the strain hardening behaviour was not needed in the reconstruction. The rectangle function has at small values for γ_0 a small amplitude and with increasing γ_0 the amplitude increases until a maximum value is reached. This course of the amplitude resembles the typical behaviour of the intensity of the third harmonic depending on γ_0. The amplitude of the sawtooth function stays always on a very low level compared with the rectangle function, note the logarithmic scale. The phases of the two functions show a small tendency to increase with increasing γ_0, but the distance between both is almost constant. The phase of the rectangle function is in the range of 360 °, meaning in-phase with the sine function, the linear contribution. After the examination of several samples this behaviour is typically found in PS-dispersions.

In contrast to this is the behaviour of the sample ckb222, that already shows an earlier onset of the non-linearities, see Fig. 9.8. At small γ_0 values the sawtooth function has a larger amplitude than the rectangle function. At a value of 0.06 for γ_0 both contributions show a strong increase. After this strong increase the amplitude of the rectangle function is at least a factor of 1.5 larger than the amplitude of the sawtooth function. A large change was also found in the phase

strain amplitude γ_0	rectangle amplitude [a.u.]	rectangle phase [°]	time-lag [ms]	sawtooth amplitude [a.u.]	sawtooth phase [°]	time-lag [ms]
0.05	0.1	337	936	0.016	239	663
0.1	0.39	338	938	0.014	217	603
0.5	1.54	341	947	0.004	206	572
1	1.74	346	961	0.004	242	672
1.5	1.72	348	967	0.01	225	625
2	1.66	361	1003	0.28	261	725
2.5	1.65	353	981	0.01	228	633
3	1.62	354	983	0.009	240	667
4	1.61	355	986	0.005	288	800

Table 9.6: The tabular shows the amplitude and the phase values of the rectangle function and of the sawtooth function. Additionally the phase lag between the linear response and the rectangle or sawtooth contribution is translated in to a time in ms. The data sets were acquired by LAOS experiments at 1 Hz from the sample ckb229. The sinusoidal response was used as reference with an amplitude of 1 and a phase angle zero.

γ_0 strain amplitude	rectangle amplitude	rectangle phase [°]	time-lag [ms]	sawtooth amplitude	sawtooth phase [°]	time-lag [ms]
0.01	0.34	139	386	0.55	319	886
0.025	0.312	141	391	0.55	329	913
0.05	0.312	138	383	0.42	331	919
0.075	1.11	-4	988	1.16	311	863
0.1	1.3	1	3	1.06	317	880
0.25	1.92	21	58	0.78	336	933
0.5	1.7	39	108	0.53	350	972
0.75	1.58	48	133	0.47	359	997
1	1.4	51	141	0.36	362	6

Table 9.7: The tabular shows the amplitude and the phase values of the rectangle function and of the sawtooth function. Additionally the phase lag between the linear response and the rectangle or sawtooth contribution is translated in to a time in ms. The data sets were acquired by LAOS experiments at 1 Hz from the sample ckb222.

of both functions at the same value for γ_0. Before this value the sawtooth function is approximately in-phase with the linear contribution, whereas the rectangle function is approximately out-of-phase (180 °). At $\gamma_0 = 0.06$ the phase of the rectangle function shows a jump, and is then approximately in-phase with the linear contribution. An interesting fact is that at this γ_0-value an overshoot can be detected in a strain sweep like in Fig. 8.7, named the Ahn type III behaviour. For a deeper insight the time domain data sets at $\gamma_0 = 0.01$, 0.1, and 1 were added to the γ_0 dependence of the amplitude, see Fig. 9.13.

With the ability to split up the time domain signal into different contributions, the possibility to separate the linear contribution and the strain softening contribution from the contribution of shear bands or wall slip should be possible. To reach the aim of separating the time domain signal into different contributions, first a reconstruction of the time domain signal was done. Then the determined contributions from the linear response and from the strain softening response, are superimposed. This superposition is based on the use of the amplitudes and phases resulting from the analysis. This reconstructed signal is then analysed with respect to the higher harmonics intensities and phases. These higher harmonics were then compared with the higher harmonics determined via the standard analysis. In the following this analysis was performed on the samples ckb229 and ckb222. The comparison of the magnitudes of the higher harmonics is shown in Fig. 9.14 and in Fig. 9.16, and the phases of the higher harmonics are shown in Fig. 9.15 and Fig. 9.17. The values of the third harmonic of the magnitudes and of the phases of the sample ckb229 can be reproduced with this new analysis. Only at small γ_0-values small deviations can be detected. After analysing several datasets of different samples this behaviour was found to be typical for PS-dispersions.

In the case of the sample ckb222 large deviations could be found in the intensity of the magnitudes of the third harmonic, see Fig. 9.16. With the standard analysis method at small γ_0-values much lower intensities for the third harmonic were found than with the analysis via the characteristic functions. As already discussed before, the rectangle function is out-of-phase at small strain amplitudes, and in-phase at larger strain amplitudes. The lower intensities of the third harmonic at small strain amplitudes, can therefore be explained by superimposing the Fourier contributions of the third harmonic originating from the

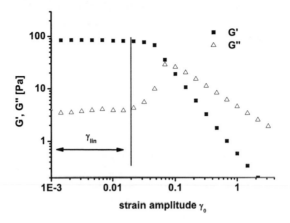

FIG. 9.7: Strain sweep plot at a frequency $\frac{\omega_1}{2\pi} = 1\ Hz$ and a temperature of $293\ K$ for dispersion CKB229. No overshoot in G' but in G'' is found and therefore this sample belongs to Ahn type III. For $\gamma_0 < 0.01\%$ the sample responds linearly. At higher strain amplitudes the sample shows strain softening.

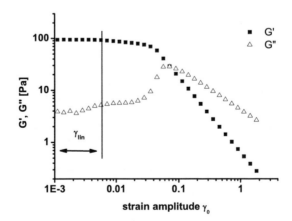

FIG. 9.8: Strain sweep plot at a frequency $\frac{\omega_1}{2\pi} = 1\ Hz$ and a temperature of $293\ K$ for dispersion CKB222. No overshoot in G' but in G'' is found and therefore this sample belongs to Ahn type III. For $\gamma_0 < 0.006\%$ the sample responds linearly. At higher strain amplitudes the sample shows strain softening.

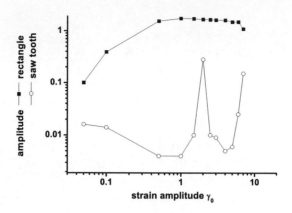

FIG. 9.9: Dependence of the amplitude on the strain amplitude of the characteristic functions of the sample ckb229. The characteristic functions are strain softening and wall slip or shear bands. Ckb229 is a typical example for the polystyrene examples.

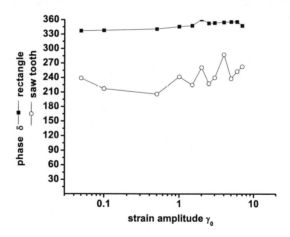

FIG. 9.10: Dependence of phase on the strain amplitude of the characteristic functions of the sample ckb229. The characteristic functions are strain softening and wall slip or shear bands. Ckb229 is a typical example for the polystyrene examples.

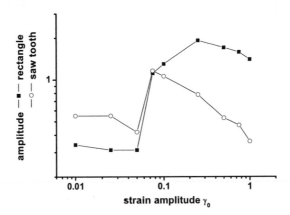

FIG. 9.11: Dependence of the amplitude on the strain amplitude of the characteristic functions of the sample ckb222. The characteristic functions are strain softening and wall slip or shear bands.

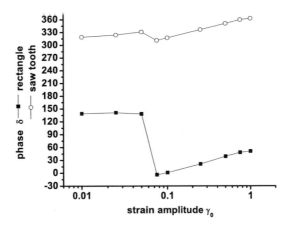

FIG. 9.12: Dependence of phase on the strain amplitude of the characteristic functions of the sample ckb222. The characteristic functions are strain softening and wall slip or shear bands.

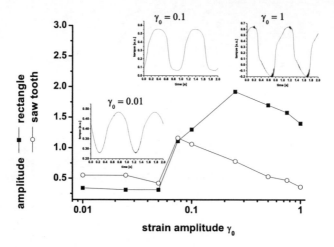

FIG. 9.13: Magnitude of the characteristic functions, rectangle function and sawtooth function depending on the strain amplitude of the sample ckb222. The plots of the time domain data at the strain amplitudes of $\gamma_0 = 0.01$, 0.1, and 1 are added.

sawtooth function and of the out-of-phase contribution of the rectangle function, see Fig. 9.18. In this plot the time domain contributions and the real part after the Fourier Transformation of the sine function, of the in-phase rectangle function, of the out-of-phase rectangle function, and the in-phase sawtooth function are shown. The sine function, and the in-phase rectangle function show only positive contributions for the harmonics, whereas the out-of-phase rectangle function shows only negative values the harmonics. The in-phase sawtooth function shows positive values for the odd harmonics and negative values for the even harmonics. In experiments the measured intensity of the third harmonic could therefore be smaller than the real existing intensity of the third harmonic, due to the reduction by the out-of-phase appearing odd harmonics in the experiment. With a different phase shift an increase of the intensity of the third harmonic, even larger than the individual contributions, could also be achieved. For a better understanding the contributions of the third harmonic from the rectangle function and the sawtooth wave originating from the equations equation (9.2) and equation (9.4) are shown in the following:

$$3^{rd}intensity = \frac{sin3[\omega t + \delta_r]}{3} + \frac{sin3[\omega t + \delta_{st}]}{3}.$$ (9.6)

The phase of the rectangle function is put to $\delta_r = 180°$, whereas the phase of the sawtooth function is out-of-phase, meaning $\delta_{st} = 0°$. This results in:

$$3^{rd}intensity = \frac{sin3[\omega t + 180°]}{3} + \frac{sin3\omega t}{3},$$ (9.7)

leading to a disappearing intensity of a third harmonic:

$$3^{rd}intensity = -\frac{sin3\omega t}{3} + \frac{sin3\omega t}{3} = 0$$ (9.8)

In the phase plots, see Fig. 9.17 differences where negligible, except for the value of the fifth harmonic at the strain amplitude of $\gamma_0 = 0.01$. This value needs further examinations. More important is the strong shift in the phases at a strain amplitude of 0.1, correlating with the strain overshoot in G' in a strain sweep.

The high non-linearities found with the new analysis method correspond to the shear thinning behaviour under steady conditions. With this method it is possible and necessary to separate the different contributions of a rheological time domain signal.

With this analysis and the ability to separate the different contributions, the time-lag (phase) between the contributions is accessible. This time-lag could be a possible access to intrinsic times. Further examination should focus on the analysis of time domain data at different frequencies.

After this method was first applied on a model system, an extension to more complex systems like polymer solutions, melts, or other rheological complex systems is desired. Possible aims are if this method can separate the non-linear response into simple, adapted functions, where the relative contributions can be used as a measure for the specific systems, and furthermore an analysis of the time-lag (phase).

For the analysis after this method a software for the reconstruction of the time domain signal was developed, based on the LabVIEW environment see Appendix B.2.

This reconstruction method was also performed on samples containing the FD-virus in chapter (10). The FD-virus is a rod-like polyelectrolyte water system,

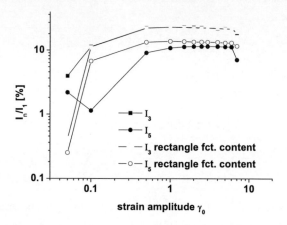

FIG. 9.14: Dependence of the intensity of the third and the fifth harmonic on the strain amplitude of the sample ckb229. The filled symbols show the results from the standard analysis, and the open symbols show the results corrected for shear band, utilising the sawtooth function

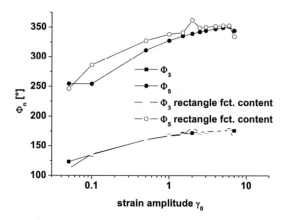

FIG. 9.15: Dependence of the phase of the third and the fifth harmonic on the strain amplitude of the sample ckb229. The filled symbols show the results from the standard analysis, and the open symbols show the results corrected for shear band, respectively the sawtooth function.

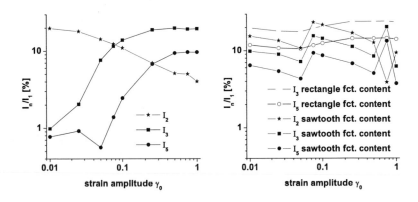

FIG. 9.16: Dependence of the intensity of the measured second, the measured third and the measured fifth harmonic on the strain amplitude of the sample ckb222 on the left side. On the right side the different contributions of the separated second, third, and fifth harmonic are shown.

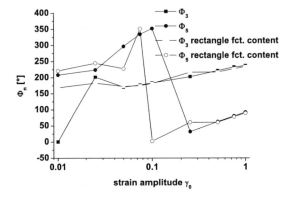

FIG. 9.17: Dependence of the phase of the third and the fifth harmonic on the strain amplitude of the sample ckb222. The filled symbols show the results from the standard analysis, and the open symbols show the results corrected for shear band, respectively the sawtooth function.

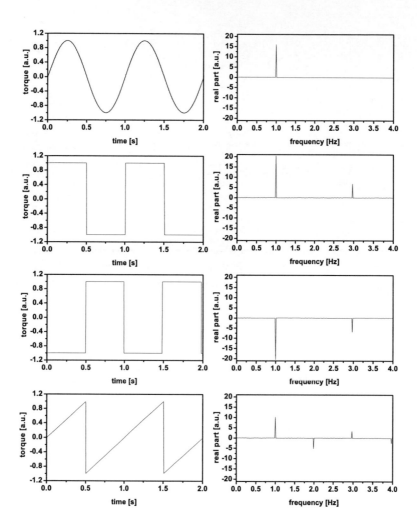

FIG. 9.18: Plots showing from the top to the bottom the time domain contributions
and the real part after the Fourier Transformation of the sine function, of the in-phase
rectangle function, of the out-of-phase rectangle function, and of the in-phase sawtooth
function. It is clearly visible that the out-of-phase rectangle function has a negative contribution of the third harmonic, whereas the in-phase sawtooth function has a positive
contribution of the third harmonic.

see chapter 5. Similar agreement between measured and reconstructed data was achieved as for the polystyrene dispersions see chapter 10.2.

9.2 Conclusion of the chapter separation of LAOS-response into characteristic response functions

In this chapter, a superposition method to analyse the frequency domain and the time domain signal for LAOS type stress responses was developed. Based on the superposition of typical non-linear response functions, like a sinusoidal, a rectangular, a triangular, and a sawtooth wave, corresponding to a linear contribution, a strain softening contribution, a strain hardening contribution and wall slip or shear bands, the time domain signal is reconstructed. Agreement with the measured time and the frequency domain data was optimised. The advantage of this approach is, that the characteristic basic functions represent known rheological phenomena. As a result, the measured signals can be described via the relative intensities and the relative phase (respective time-lag) to each other. This new approach was tested for the two model dispersions as very accurate and promising. In a next step this analysis was then applied to a set of time domain signals with increasing strain amplitudes. Here a typical behaviour for the polystyrene dispersions was found to be that the phase of the rectangle function is in-phase with the linear contribution, and that the magnitude shows a behaviour similar to the third harmonic in the standard analysis.

In the standard analysis the time domain signal is Fourier transformed and afterwards the relative intensity $\frac{I_3}{I_1}$ and the phase of the third harmonic Φ is analysed. In cases, where the Fourier spectrum exhibits a significant number of higher harmonics, as well as even harmonics in addition to the odd ones normally observed, this analysis is not adequate. Considering the whole spectrum as resulting from a separation of characteristic responses a separation of the sine function contribution and rectangle function contribution from the sawtooth function contribution is more adequate and was applied to a sample with a large amplitude in the sawtooth function. After the separation the analysis showed much larger values for the intensity of the third harmonic at small strain amplitudes, whereas the values of the phase were like in the standard analysis. The reason therefore is the appearing out-of-phase rectangle function with negative values of the third harmonic intensity, and the positive values of the third harmonics intensity of the sawtooth function. So smaller intensities of the third harmonic are actually measured, than that are in the sample due to interfering intensities of the rectangle function and the sawtooth function. Thus, this method is indeed able to separate the different contributions, and is recommended to analyse samples where a large second harmonic is found.

Chapter 10

Rheological behaviour of rod-like particles

This chapter focusses on the mechanical analysis of samples containing the FD-virus. The FD-virus is a monodisperse biological system with a length of ≈ 880 nm, a diameter of 20 nm, and a persistence length of $2,200$ nm. Several concentrations of the FD-virus dispersions have been analysed. These concentrations range from the overlap concentration c^* up to 300 c^*. The concentration is determined according to the procedure mentioned in chapter 5.2, with the overlap concentration for the FD-virus of 0.04 $\frac{mg}{ml}$. Typical results for the measurements will be presented in the following.

10.1 Non-linear rheological properties of the FD-virus

10.1.1 Non-linear behaviour of FD-virus dispersions under strain dependent LAOS conditions

Complex fluids, like watery dispersions containing the FD-virus, show a great variety of rheological properties. Under oscillatory shear several different types of behaviour are known. Ahn et al. have conducted oscillatory strain sweep tests on several different materials. They found four different classes of behaviour under strain sweep tests [Hyun 02, Sim 03] in their experiments (see Fig. 8.5), which

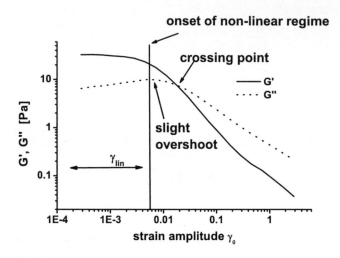

FIG. 10.1: The strain dependence of the storage G' and loss G'' modulus of a sample containing the FD-virus at a frequency of $\frac{\omega_1}{2\pi} = 1\ Hz$ and at a temperature of $298\ K$. At a concentrations of $300\ c^* = 12\ \frac{mg}{ml}$ a behaviour of type III according to Ahn is found.

were introduced and described in chapter 8.2.1. Only type III will be treated here, because no other type was found in samples containing the FD-virus. This type III is characterised by a strain overshoot in G'' [Hyun 02]. Ahn concluded this response for systems with a weak network-like structure. During the application of a shear field this network-like structure resists first the deformation up to a certain strain. As a result the modulus G'' increases. After reaching a critical deformation these network-like structures are destroyed, and both moduli decrease. A typical example is given by a water-based dispersions containing the FD-virus at a concentration of $300\ c^*$ ($1\ c^* = 0.04\ \frac{mg}{ml}$) (see Fig. 10.1). A type III was found, but with a very small overshoot, when comparing to the dispersions synthesised via mini-emulsion polymerisation. This can be explained by the interactions between the viruses due to its rod-like structure. Therefore the resistance of the structure against deformation seems to be much smaller than in the case of the dispersions, and therefore leads to a smaller overshoot.

10.1.2   Mechanical analysis of the FD-virus via FT-rheology

Due to its relative simple and defined shape and the water solubility the FD-virus is a very interesting model system for the FT-rheology. The FT-rheological measurements are performed and analysed like the once described in chapter 8.2.2. Oscillatory measurements at a constant frequency and succeeding change in strain amplitudes were performed. Additionally the shear rate dependent viscosity η was determined (see Fig. 10.2). For a concentration of $300\ c^*$ the non-linear measurements are performed and analysed with respect to the appearance of higher harmonics, and for the prediction of the higher harmonics the Carreau parameter were determined from the shear rate dependent viscosity η. Afterwards the prediction of the higher harmonics, based on the Carreau parameters, was conducted.

In the analysis of the magnitudes of $\frac{I_3}{I_1}$, $\frac{I_5}{I_1}$... (see Fig. 10.3), it is clearly visible that the predictions at large strain amplitudes overlay well with the measured values, but are less intense at smaller strain amplitudes. This is a striking difference compared to the results known from the dispersions, where the predic-

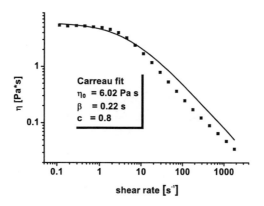

FIG. 10.2: The shear rate dependent viscosity η of a sample containing FD-virus at a concentration of $300\ c^*$ and at a temperature of $298\ K$. The parameters extracted via the Carreau model are defined as: $\eta_0 = 6.02\ Pas$, $\beta = 0.22\ s$, $c = 0.8$.

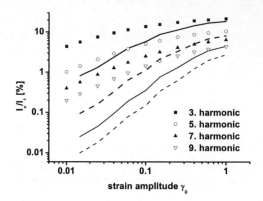

FIG. 10.3: Comparison of measured and predicted higher harmonics magnitudes for FD-virus at a concentration of 300 c^*. The applied frequency was $\frac{\omega_1}{2\pi} = 1\ Hz$ at a temperature of 293 K. The Carreau parameters extracted from Fig. 10.2 are: $\eta_0 = 6.02\ Pas$, $\beta = 0.22$ s, $c = 0.8$. The lines are the predicted values.

tions are normally larger at small strain amplitudes, and the fit is quite good for larger strain amplitudes (see Fig. 8.15 in chapter 8.2.2). The phases of the 3rd and 5th harmonics are constant around the values of $195 - 205$ ° and $27 - 33$ ° (see Fig. 10.4). In shear rate dependent viscosity measurements shear thinning behaviour was found (see Fig. 10.2). According to Neidhöfer [Neidhöfer 03a], values of 180 ° an 0 ° for the higher harmonics phases were expected for strain softening under oscillatory shear, which is the equivalent effect to shear thinning under steady shear. A larger offset of 10 ° up to 50 ° seen in the 3rd and 5th phase is observed in dispersions synthesised via mini-emulsion polymerisation also.

Furthermore the intensity of the 2nd harmonic is larger than $3\ \%$ (see Fig. 10.5) over the observed span of strain amplitudes, and can therefore not be neglected. The appearing phases of the 2nd-harmonics are stable and reproducible (see Fig. 10.6).

The appearance of the second harmonic might be explained on the appearance of shear bands or breaking up of aggregates. In the work of Graham, Heymann and Ahn [Graham 95, Heymann 01, Sim 03] wall slip or breaking up of aggre-

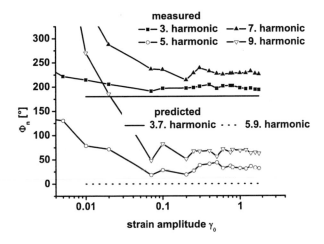

FIG. 10.4: Comparison of measured and predicted odd higher harmonics magnitudes for FD-virus at a concentration of 300 c^*, corresponding to Fig. 10.3.

gates does generate even harmonics. The difference between the calculations and the dramatic increase of the intensity of the 3rd and 5th at very small strain amplitudes, could be based on the fact, that the assumptions done for the prediction based on the Carreau model are not correct. The assumption of a purely viscous response can be proved to correct by the frequency dependence (see Fig. 10.7) of G' and G''. It may not be correct to assume an instantaneous adjustment at small strain amplitude, which would hint on internal forces like weak networks. Results showing a weak network were already found in the strain dependence of G'' (see Fig. 10.1). The internal processes like aligning and tumbling of the rod-like FD-virus under the change of shear direction could be the reason for the phase lag. Additionally the 'bulk' liquid crystal could break up into layers, which might glide along each other. This breakup should occur along the director of the particles in the nematic liquid crystal. This gliding could be interpreted as shear bands. But likewise in the case of the dispersions no theory is available yet, that could explain the appearance of the second harmonic in combination

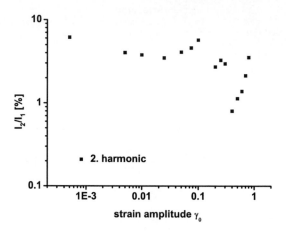

FIG. 10.5: Magnitude of the second harmonic measured at a concentration 300 c^*, corresponding to Fig. 10.3.

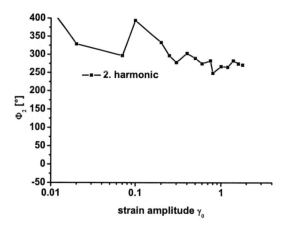

FIG. 10.6: Measured phase of the second harmonic for a FD-virus dispersion of a concentration of 300 $c^* = 12 \frac{mg}{ml}$, corresponding to Fig. 10.3.

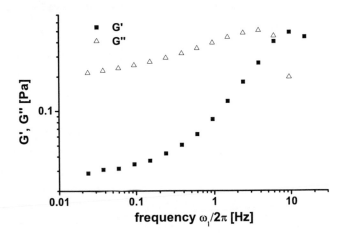

FIG. 10.7: The frequency dependence of the storage G' and loss G'' modulus of a sample containing the FD-virus of a concentration of 300 $c^* = 12 \frac{mg}{ml}$ and at a strain amplitude of $\gamma_0 = 0.01$. A dominantly viscous response is found for low frequencies.

with the offset found in the odd phases. Only a similar result is known for wall slip [Graham 95, Sim 03, Heymann 01], which would be a reasonable model for shear bands, that are based on the same effect, that two objects glide along each other.

10.1.3 Prediction of non-linear rheological properties according to Dhont

In chapter 2.4.1 a theory to predict the non-linear rheological behaviour of rod-like particles is presented. This theory developed by Dhont et al. [Dhont 03] was applied on FD-virus dispersions. Non-linear dynamic measurements were performed on the samples containing FD-virus to check the validity of this theory.

In Fig. 10.8 the theoretical concentration ($\frac{L}{d}\varphi$) dependence of the η_{eff}/η_0 is plotted for several rotational Péclet-numbers ($Pe_r = \frac{\gamma_0}{D_r}$), given by the equation (2.83). The η_{eff}/η_0 shows a strong increase with a maximum in the concentration dependence. This increase is larger for smaller Péclet-numbers, and they also have higher absolute values. For higher concentrations these curves then reach a constant value. In Fig. 10.9 the analysis of the performed measurements

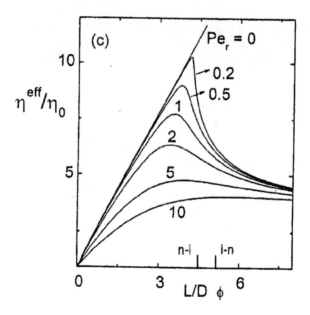

FIG. 10.8: Calculation based on the theory of Dhont [Dhont 03] of the concentration dependence of η_{eff}/η_0 given at different shear rates for Pe_r-number values from 0 up to 8.

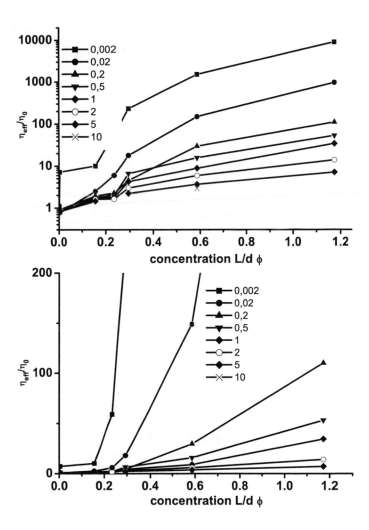

FIG. 10.9: Analysis of the concentration dependence of η_{eff}/η_0 given at different Pe_r-number values from 0.002 up to 10. The two plots show the same data, but the lower one just shows the part of small values for η_{eff} to show the strong increase, for small Péclet numbers of 0.002 and 0.02.

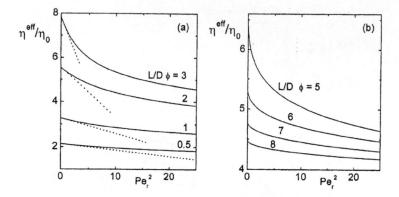

FIG. 10.10: Theoretical suspension viscosity η_{eff} normalised with the solvent shear viscosity η_0 as function of squared Péclet number (Pe_r^2) for several concentrations $\frac{L}{d}\varphi$. Whereas the graph a) shows the isotropic and b) the nematic state [Dhont 03]. Note, the concentration in the plot is given by $\frac{L}{d}\varphi$.

show the same trends in the behaviour as given by the theoretical predictions, but the increase is stronger in the measurements than in the predictions. Additionally much larger values for η_{eff}/η_0 are found, but the same trends can be seen, in particular, if the lower plot from Fig. 10.9 is compared with Fig. 10.8.

In Fig. 10.10 the prediction for the suspension viscosity η_{eff}/η_0 is plotted as a function of the squared Péclet number (Pe_r^2) for several concentrations $\frac{L}{d}\varphi$ ranging from 0.5 up to 3 in the isotropic state and ranging from 5 to 8 in the nematic state. The dotted lines correspond to the low shear-rate expansion equation (2.82). These theoretical calculations are done on the basis of equation (2.82) [Dhont 03]. The η_{eff}/η_0-values are higher the higher the concentration of the examined dispersions become. For small values of Pe_r^2 approximating 0 an increase is found. The higher the concentration the higher is also the slope. For increasing Pe_r^2-numbers the different concentrations show a asymptotic behaviour. Analysis of rheological measurements on FD-virus dispersions at concentrations ranging from 0.1 c^* up to 300 c^* are given in Fig. 10.11. Similar trends are found for the measured and the predicted data, but η_{eff}/η_0-values of the measurements are about 10 times higher than the predicted values.

It may be concluded that the theory of Dhont et al. results in calculations of

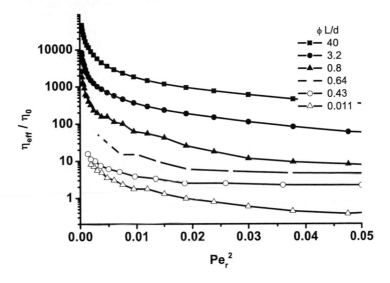

FIG. 10.11: Suspension viscosity η_{eff} normalised with the solvent shear viscosity η_0 as a function of the squared rotational Péclet number for several concentrations $\frac{L}{d}\varphi$ (40, 3.2, 0.8, 0.64, 0.43, 0.011) at a frequency $\frac{\omega_1}{2\pi} = 1\,Hz$.

the FD-virus dispersions, that can reproduce the same trends in the prediction compared with the actual behaviour, but with deviations in the absolute values.

10.1.4 Non-linear dissipative and elastic contribution of the viscosity

In the theory developed by Dhont et al. [Dhont 03] equations for the calculation of the non-linear behaviour of the higher harmonics were derived. Based on the equations for the viscous contribution (equation (2.89)) and for the elastic contribution (equation (2.90)), introduced earlier (see chapter 2.4.1), the magnitudes and phases of the third and the fifth harmonic have been calculated and are presented in Fig. 10.12. The intensities and phases are valid for a frequency of 3.3 Hz at a concentration of $c = 25.5$ c^* in the upper part, and a concentration at $c = 76.6$ c^* in the lower part of the Fig. 10.12. These data sets were kindly provided by Dr. P. Lettinga in the group of Prof. J. Dhont.

The magnitudes of the third and fifth harmonic show larger non-linearities for higher concentrated samples. The maximum non-linearity is 8 % intensity for the third harmonic, at a concentration of $c = 25.5$ c^* and 10 % at a concentration of $c = 76.6$ c^*. The phases of the third and the fifth harmonic approximate values of 320 ° respective 280 ° for the lower concentrated sample and values of 280 ° respectively 220 ° for the higher concentrated sample.

In Fig. 10.13 the results of the measurements of the higher harmonics magnitudes and phases of a sample containing FD-virus are shown. The frequency is 2.5 Hz. The measurements are performed at concentrations of $c = 25$ c^* in the upper part, and $c = 60$ c^* in the lower part and a temperature of 25 °C. For both frequencies very high non-linearities are found with magnitudes of above 25 % at strain amplitudes of above $\gamma_0 = 0.5$. For increasing strain amplitudes the intensities level off. The magnitudes at a frequency of 2.5 Hz have higher intensities than those at frequencies of 6 Hz. The phases of the measurements at 2.5 Hz reach the expected values of 0 ° for the third harmonic and 180 ° for the fifth harmonic already at smaller strain amplitudes γ_0.

The calculations show, like the measurements, an increase of the magnitudes of the higher harmonics, but these magnitudes have a three times lower intensity. Additionally the intensity is predicted to have a larger value for the higher concentration, but the opposite is found in the measurements. Second harmonics, that were found in the measurements see Fig. 10.5 and Fig. 10.6, were not predicted by the calculations based on the theory of Dhont. The phases of the third

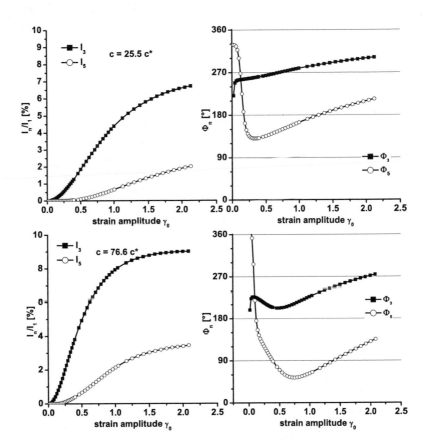

FIG. 10.12: Theoretical predictions of the non-linear mechanical behaviour of a sample containing FD-virus, based on the theory of Dhont (see chapter 2.4.1) are shown here. The magnitudes and phases of the third and the fifth harmonic are plotted for a frequency of 3.3 Hz at the concentrations of $c = 25.5\ c^*$ in the upper part, and $c = 76.6\ c^*$ in the lower part.

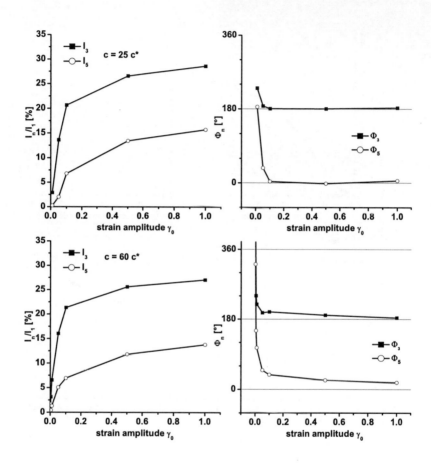

FIG. 10.13: Results of the non-linear rheological measurements of a sample containing FD-virus. The magnitudes and phases of the third and the fifth harmonics are shown here. The frequency was $2.5\ Hz$ and the measurements were done at concentration of $c = 25\ c^*$ in the upper part, and a concentration of $c = 60\ c^*$ in the lower part.

and the fifth harmonic approximate limit values for larger strain amplitudes γ_0, both in measurements and calculations, but these limit values are different. In the case of the measurements the limiting value for the third harmonic is 180 ° in the experiments compared with 320 ° and 220 ° resulting for the different concentrations from the calculations. For the fifth harmonic similar results are found. The limiting value in the experiments was found to be 0 ° and 280 ° or 220 ° originating from the calculations. The calculations are able to predict the same trends in the behaviour of the higher harmonics magnitudes and phases, but show large deviations in the absolute numbers. An explanation for the lower measured intensities at a concentration of $c = 60\ c^*$ than the predicted ones, might be the cross-over from the isotropic to the nematic regime which is close to this concentration.

Another hint on the ability to predict the same trends between the predictions and the measured behaviour of the magnitude of the third harmonic is also found while fitting the data with equation (2.69). This function, developed by Wilhelm [Wilhelm 02] to describe the strain dependence of the magnitude of the third harmonic, describes the predicted values and overlays with them qualitatively and quantitatively, see Fig. 10.14. This is only a proof for a qualitative agreement of the theory and the measurements because the parameter A which describes the maximal reachable intensity of the third harmonic is a free parameter and therefore the quantitative agreement is not proofed via this method.

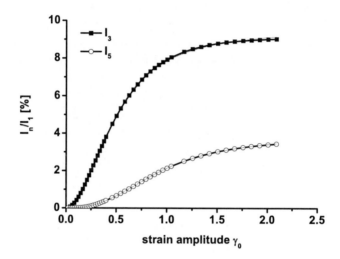

FIG. 10.14: Theoretical predictions of the non-linear mechanical behaviour of a sample containing FD-virus, based on the theory of Dhont (see chapter 2.4.1) are shown here. The magnitudes and phases of the third and the fifth harmonic are plotted for a frequency of 3.3 Hz at the concentration of $c = 76.6\ c^*$. This data is fitted with equation (2.69) that describes the strain amplitude dependence of the 3rd-harmonic.

10.2 Separation of LAOS-response into characteristic response functions applied on a FD-virus dispersion

After introducing the new analysis method of the non-linear oscillatory rheological data in chapter 9, this analysis was also applied to samples containing the FD-virus. In the following the results are presented.

The reconstruction of the time domain data was performed at different strain amplitudes ranging from $\gamma_0 = 0.1$ up to 2. The reconstruction could be achieved, as introduced in chapter 9. The characteristic functions needed for the reconstruction were sine (linear contribution), rectangle wave (strain softening contribution), and sawtooth wave (shear band or wall slip contribution). As can be seen in Fig. 10.15 the amplitude of the rectangle wave is normally at least double the amount of the amplitude of the sawtooth wave. The shear bands or the wall slip should therefore show almost no influence on the values of the higher harmonics intensities. The phases of the rectangle wave, and the sawtooth wave, see Fig. 10.16, show both a value of about 30 ° to 50 °. The strain softening process is therefore at the same time like the linear contribution of the response with respect to the excitation. The further analysis of these data sets reveals that the intensities and the phases of the third harmonic originating from the standard analysis and from the analysis with characteristic functions overlay well, see Fig. 10.17, and Fig. 10.18. The differences found in the magnitude and phase of the fifth harmonic at smaller strain amplitudes have their origin in the use of rectangle wave, with different intensities in the fifth harmonic. The analysis should now be applied to other concentrations to see if this analysis could be applied there too.

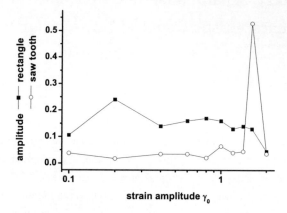

FIG. 10.15: This plot shows the dependence of the amplitude on the strain amplitude of the characteristic function of the sample FD-virus at a concentration of $C = 0.1\ c^*$. The characteristic functions are strain softening and wall slip or shear bands. It also includes the intensity of the third harmonic depending on the strain amplitude.

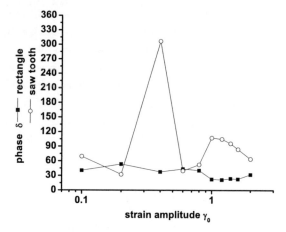

FIG. 10.16: This plot shows the dependence of phase on the strain amplitude of the characteristic function of the sample FD-virus at a concentration of $C = 0.1\ c^*$. The characteristic functions are strain softening and wall slip or shear bands.

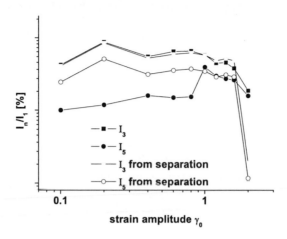

FIG. 10.17: This plot shows the dependence of the intensity of the third harmonic on the strain amplitude of the sample ckb222. The filled symbols show the results from the standard analysis, and the open symbols show the results corrected for shear band, respectively the sawtooth function.

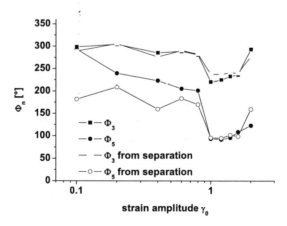

FIG. 10.18: This plot shows the dependence of the phase of the third harmonic on the strain amplitude of the sample ckb222. The filled symbols show the results from the standard analysis, and the open symbols show the results corrected for shear band, respectively the sawtooth function.

10.3 Conclusion of chapter the rheological behaviour of rod-like particles

In this chapter the non-linear mechanical behaviour of dispersions containing the FD-virus were examined. In general those samples show very high non-linearities at already very low volume fractions, when compared to polystyrene dispersions. Likewise, the polystyrene dispersions, calculations of the build-up of the higher harmonics were conducted. These calculations cannot predict the experimentally detected strong increase of the magnitudes $\frac{I_3}{I_1}$, $\frac{I_5}{I_1}$ at small strain amplitudes, but give reasonable results for larger strain amplitudes. Additionally the second harmonics, found in the measurements, were not predicted by the calculations. A possible explanation for the second harmonics are shear bands, or wall slip. Predictions for the phases Φ_n give an offset similar to the one found in dispersions synthesised via mini-emulsion polymerisation. These phase values can originate from shear bands, but this has to be proven by more demanding calculations. Note, the volume fraction in the case of the FD-virus is below $3\ vol.\%$.

Additionally comparisons of measurements with calculations based on the theory developed by Dhont et al. [Dhont 03] revealed the same trends to be found in η_{eff}/η_0 as a function of the concentration ($\frac{L}{d}\varphi$) at several Péclet-numbers (Pe_r) and the suspension viscosity η_{eff}/η_0 as a function of the squared Péclet number (Pe_r^2). Furthermore calculations have been conducted to describe the non-linear behaviour utilising I_3/I_1, I_5/I_1, Φ_3, and Φ_5. These calculations also showed the same trends, but higher intensities are found in the measurements than in the calculations. The calculation of the phases Φ_n of the higher harmonics shows much lower values than those found in the measurements. Furthermore the even harmonics were not predicted by the theory of Dhont.

The new analysis method for oscillatory rheological data was performed on a FD-virus solution with a concentration of $c = 1\ c^*$. It was possible to reconstruct the time domain signal at different strain amplitudes with the help of the characteristic functions sine, rectangle, and sawtooth. Furthermore the intensities and the phases of the third harmonic originating from the standard analysis and from the analysis via characteristic functions overlay well.

Chapter 11

Summary and Outlook

Within this thesis several aspects of non-linear mechanical shear on water-based systems were investigated, mostly using FT-rheology. In the first step polystyrene dispersions with a high solid content were synthesised by using the emulsion polymerisation technique. Two different synthetic pathways were chosen, the semi-continuous emulsion polymerisation and the mini-emulsion polymerisation. The polystyrene dispersions were characterised with regard to their solid content, stability, particle size, and particle size distribution. Mechanical analysis was conducted by using rheology in the linear and non-linear regime and especially FT-rheology. Additionally, a new method to analyse the measured time domain signal and the frequency domain signal, via characteristic functions was developed. Further complex fluids, such as the biological system FD-virus and a concentrated polystyrene solution in di-*sec*-octyl phthalate (DOP), were mechanically characterised and used to further develop the FT-rheo-optical technique. The FT-rheology was extended to the area of rheo-optics. The last topic of the results concerns the mineralisation of ZnO under the influence of polymer addition when a defined shear field is applied.

The rheo-optical set-up, especially the background correction of the birefringence, was improved within this research. The measured rheo-optical signal is strongly influenced by the background birefringence. The negative influence of the background birefringence is reduced by the factor of 20 via a variable retarder, application of an oversampling technique and additional mathematical treatment. The results obtained are confirmed via theoretical calculations based on a rotating birefringent object. It is now possible to explain and simulate the effect of a background birefringence vector. Additionally, these modifications increase the resolution in the time domain by the factor of 24. With the improvements of the rheo-optical set-up, a tool combining rheo-optic and FT-rheology is now available. With this set-up, non-linearities can be examined simultaneously with rheology and with birefringence on the one hand or with dichroism on the other.

The influence of polymers under a defined shear field on the mineralisation process of inorganic ZnO was examined. The effect of shear on the crystallisation process was analysed by comparing the size, the shape, and the aspect ratio of the crystallites. The examination of different polymers at different shear rates confirms the already known change in the aspect ratio. Additionally, a change of the size of the particles is observed. The polymer with the structure P(EO-co-AA) causes a significant reduction in both diameter and length by the factor of 2. At higher shear rates abrasive effects are detected. The edges are less sharp and fewer twin crystals are found. The ability to reduce the crystal size by the factor of 2, offers a pathway to easily modify crystals by the addition of polymer in combination with a defined shear field. Noticeable effects are only visible when a polymer with a high molecular weight is used. Therefore, it can be concluded that the shear has only an effect on the polymers with a sufficiently high molecular weight.

Two different types of polystyrene dispersions with very high solid content (each $\varphi > 0.3$) were examined especially by using LAOS experiments. The differences in the synthesis of the dispersions cause variations in the amount of ions in the solution, in the polydispersity and in the Debye-length. The mechanical behaviour of these dispersions varies in the strain dependence of G' and G'' at the onset of the non-linear regime. An overshoot in G'' is detected only in dispersions synthesised via mini-emulsion polymerisation and not in those synthesised via semi-continuous emulsion polymerisation, which is explained by the lower ion loading and therefore higher Debye-length in the case of dispersions synthesised via mini-emulsion polymerisation ($1.8 \ nm$ in the case of mini-emulsion polymerisation and $0.3 \ nm - 0.4 \ nm$ in the case of semi-continuous emulsion polymerisation) and lower polydispersity (0.01 in the case of mini-emulsion polymerisation and 0.3 in the case of semi-continuous polymerisation). This effect is also observed in the FD-virus dispersions, a biological model system, but with a less intense overshoot.

Differences between the two differently synthesised dispersions are also observed via the FT-rheological analysis. For both types of polystyrene dispersions mechanical higher odd harmonics can be detected and predicted via a simplified model. In this model, instantaneous adjustment to the applied shear, and no memory in the observed time span are assumed. In dispersions, synthesised via mini-emulsion polymerisation, second harmonics were additionally found. These even harmonics are not predicted by the above mentioned model. A possible explanation for even harmonics are shear bands.

Furthermore, a method was developed to reconstruct the measured time domain signal via four different basic functions, that correspond to rheological phenomena. The reconstructed time domain data overlays well with the measured time domain data for the here examined polystyrene dispersions. In a next step this analysis was then applied to a set of time domain signals with increasing strain amplitudes. Here a typical behaviour for the PS-dispersions was found to be that the rectangle wave is in phase with the applied sine wave, and that the magnitudes show a behaviour similar to that of the third harmonic in the standard analysis. Additionally a separation of the signal in a sine wave, and a rectangle wave from the sawtooth wave was achieved. The FT-analysis of the resulting data showed similar results for the intensity of the third harmonic in the both methods of analysis, but a change in the regime where the sawtooth wave was dominating the rectangle wave. A separation of the time domain signal could thus be achieved. It was found that the Fourier contributions from the in-phase sawtooth function and the out-of-phase rectangle function have different signs and therefore reduce the intensity of the measured odd harmonics.

Samples containing FD-virus mechanically respond with very high non-linearities at a very low volume fraction ($1 < \%\ wt.$). Calculations of the build-up of the higher harmonics were conducted. These results agree with the experiments at higher strain amplitudes. These calculations cannot predict the strong increase of the higher harmonics at very low shear rates. Comparing the experiments with the prediction based on the theory developed by Dhont et al. [Dhont 03] the same trends for the suspension viscosity η_{eff}/η_0 as a function of the concentration ($\frac{L}{D}\varphi$) at several Péclet-numbers (Pe_r) and the suspension viscosity as a function of the Péclet number were found. Furthermore, calculations were conducted to describe the non-linear behaviour with the help of magnitudes and phases of the third harmonic and the fifth harmonic. These calculations show the same trends in the behaviour, but higher intensities are found in the measurements. In the phases of the higher harmonics much lower values are found in the measurements. The new analysis method via characteristic functions was applied to a FD-virus dispersion. The time domain data could be reconstructed and a separation of a sawtooth contribution was successful.

The detection of even harmonics in the non-linear rheological analysis of dispersions demands the development of a suitable theory in the standard analysis. Future research might be concentrated on the examination of the reconstruction of the measured signal, with a special focus on a better understanding of the influence of shear bands on the shear thinning process. In general, this method should be applied to samples like polymer melts and polymer solutions with an additional focus on the behaviour of the magnitudes and phases of the higher harmonics. A special focus should be on the phase (time lag) between the different characteristic functions, especially under a change of the excitation frequencies. These could represent intrinsic (relaxation) processes. After the improvement of the Rheo-optical set-up, an extension of the non-linear mechanical analysis to birefringence and dichroism, and a further extension of the non-linear mechanical analysis to the low torque regime seems to be possible and is desired.

Appendix A

Rheo-Optics

A.1 Background compensation for small angles

Two birefringent elements with retardation o and p, at angles ϵ and ι are equivalent to a single birefringent element with combined retardation q at an angle φ equation (A.1) where:

$$q^{(i2\varphi)} = o^{(i2\epsilon)} + p^{(i2\iota)}.$$ (A.1)

This can be shown by using a retardation o, p that are so small that we only have to retain terms of first order in o, p in the Mueller matrices [Fuller 95]. From the Mueller matrices using the notation $s_{2\epsilon} = sin2\epsilon, c_{2\epsilon} = cos2\epsilon$ follows:

$$M(p, \epsilon) = \begin{pmatrix} 1 & 0 & 0 & 0 \\ 0 & 1 & 0 & -os_{2\epsilon} \\ 0 & 0 & 1 & os_{2\epsilon} \\ 0 & os_{2\epsilon} & -oc_{2\epsilon} & 1 \end{pmatrix},$$ (A.2)

$$M(p, \theta) = \begin{pmatrix} 1 & 0 & 0 & 0 \\ 0 & 1 & 0 & -ps_{2\theta} \\ 0 & 0 & 1 & ps_{2\theta} \\ 0 & ps_{2\theta} & -pc_{2\theta} & 1 \end{pmatrix},$$ (A.3)

and:

$$M(q, \varphi) = M(p, \epsilon)M(p, \theta) =$$

$$M(q, \varphi) = \begin{pmatrix} 1 & 0 & 0 & 0 \\ 0 & 1 & 0 & -os_{2\epsilon} + ps_{2\theta} \\ 0 & 0 & 1 & os_{2\epsilon} + ps_{2\theta} \\ 0 & os_{2\epsilon} + ps_{2\theta} & -(oc_{2\epsilon} + pc_{2\theta}) & 1 \end{pmatrix}. \tag{A.4}$$

From this we find the already presented identities:

$$qs_{2\varphi} = os_{2\epsilon} + ps_{2\iota}, \tag{A.5}$$

$$qc_{2\varphi} = oc_{2\epsilon} + pc_{2\iota}, \tag{A.6}$$

i.e.

$$q^{(i2\varphi)} = o^{(i2\epsilon)} + p^{(i2\iota)}.$$

A.2   General background compensation

The result of A.1 is only valid if both retardations are small. In general only the background birefringence is only small after pre-compensation using a variable retarder (see section 6.2.5 on page 83 and Fig. 6.7). We represent the background birefringence as a virtual birefringence element in front on the sample. The following matrix equation (A.1) shall represent this background birefringent element [Fuller 95]:

$$M(p, \theta) = \begin{pmatrix} 1 & 0 & 0 & 0 \\ 0 & 1 & 0 & -ps_{2\theta} \\ 0 & 0 & 1 & ps_{2\theta} \\ 0 & ps_{2\theta} & -pc_{2\theta} & 1 \end{pmatrix}. \tag{A.7}$$

We used that the retardation p of the background is assumed to be so small by the variable retarder that $\sin\theta \approx \theta$. Incorporating this virtual birefringent element in the optical train we find after using the Mueller matrices and equating the complex and real parts:

$$\sin\delta'_t \sin(2\theta_t) = \sin\delta'_m \sin(2\theta_m) \pm p\cos\delta'_t \sin(2\theta), \tag{A.8}$$

$$\sin\delta'_t \cos(2\theta_t) = \sin\delta'_m \cos(2\theta_m) \pm p\cos\delta'_t \cos(2\theta). \tag{A.9}$$

The index t indicates the true signal, the index m the measured signal and p and θ the background retardation and angle. The uncertainty of the prefactor ($\pm$) is caused by the ambiguity with respect to the orientation of the birefringence of the background signal. For small retardation this results reduces to the result in A.1.

A.3 Calibration

A explanation on the calibration of the orientation angle θ is given in this chapter. A linear polariser oriented at 45 ° degrees gives an equivalent signal to a quarter-wave plate at zero degrees. The orientation angle may be set at zero and $\sin\delta$ = 1 with a linear polariser at 45 °. Here it is important to note that both, the linear polarised and the quarter-wave plate have different Mueller matrices, but the detected signal is the same [Fuller 95], since:

$$I = I_0[1+(c_{2\theta}\tanh\delta''+s_{2\theta}\sin\delta')\cos(4\omega_1 t)-(s_{2\theta}\tanh\delta''+c_{2\theta}\sin\delta')\sin(4\omega_1 t)], \quad (A.10)$$

for our set-up with an optical element having the complex retardation $\delta = \delta' + i\delta''$ at angle θ. For a quarter-wave plate, using $\delta' = \frac{\pi}{2}$ and $\delta'' = 0$ equation (A.10) can be simplified to:

$$I = I_0[1 + s_{2\theta,quarter-waveplate}\cos(4\omega_1 t) - c_{2\theta,quarter-waveplate}\sin(4\omega_1 t)]. \quad (A.11)$$

For a linear polariser, using $\delta' = 0$ and $\delta' = \infty$, equation (A.10) can be simplified to:

$$I = I_0[1 + c_{2\theta,polariser}\cos(4\omega_1 t) - s_{2\theta,polariser}\sin(4\omega_1 t)]. \quad (A.12)$$

So, for $\theta_{quarter-waveplate} = 0$ we have to choose $\theta_{polariser} = \frac{\pi}{4} + n\pi$ to get an equivalent signal.

Appendix B

LabVIEW

The LabVIEW^{TM} VI FT-Measurement [LabVIEW 98] (VI means virtual instrument) and the VI basic-analyser, used for the Fourier transformation of the acquired data, are already described elsewhere [Neidhöfer 03a]. In this Appendix the VI basic-analyser, see chapter B.1, with the extended window function and the Rheo-Optic software (VI Birefringence-Dichroism see chapter B.3) are presented and described.

B.1 VI Analyser with extended window function

The VI Analyser Fourier transforms the data acquired by the VI FT-Measurement [Neidhöfer 03a]. This special version presented here (see Fig. B.1) has as an add-on the possibility to select only a part from the acquired data-set. The software is started by pressing the arrow below the menu line. By pressing the 'open new file' button the file for analysis is selected. In this mode one can change the window size or the channel and the graphs will automatically adjust. The saving of the results is done with the use of the button 'SAVE'. By pressing the 'STOP' button the analysis is stopped. Now a new file is selected by pressing the 'open new file' button. With the 'CLOSE' button the program is stopped. In the upper right corner the complete data-set of the imported file is visible. Directly beneath the window the selected part of the data-set is shown. The size of this window is selected via the two controls on the left side in the control panel box called: 'Window begin' and 'Window end'. The default value is zero in both controls, which means that automatically the complete data file is used. By insertion of the number of

the data points, taken from the upper right graph, just the selected part is visible
in the lower right graph and is then analysed. The two big graphs on the left
side show the analysis results. The upper graph shows the magnitude spectra
and the lower graph shows the phase spectra of the Fourier transformed data set.
The values of the harmonic magnitudes and phases are shown in the indication
panel on the lower right. Furthermore the used data file, the array length of the
selected window, the file length, the Nyquist frequency and the sampling rate of
the recording are also indicated here. Additionally δ and $tan\delta$ are also shown.
In the case that several recorded channels are required, they are selected via the
channel control in the control panel. In Fig. B.2 the structure of the software is
shown.

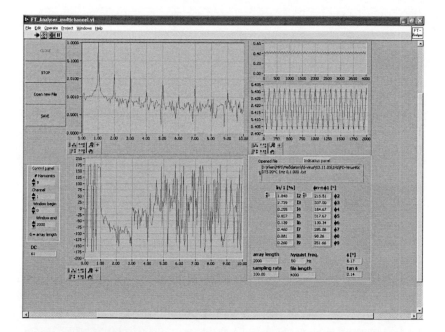

FIG. B.1: Front panel of the virtual instrument 'basic-analyser' with window function.

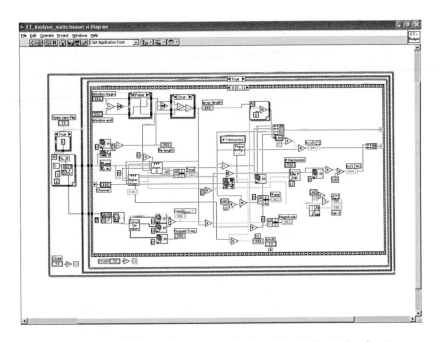

FIG. B.2: Diagram of the virtual instrument 'basic-analyser' with window function.

B.2   VI Reconstruct time signal

With the VI Reconstruct signal, the analysis of the time domain signal is done with respect to the characteristic functions: sinusoidal, rectangular, triangular, and sawtooth. This software has two different modes, an simulation and an overlay mode. In the simulation mode an addition of the different contributions can be simulated. Free parameters are the number of points and the sampling rate in $[1/s]$. In the overlay mode an measured data set can be read from a file and then overlayed with the reconstructed signal. The channel from the file can be freely selected. The data from the file and the simulated or reconstructed data is Fourier analysed and the magnitudes and the phases of the higher harmonics are separately presented. The amplitude, the phase, and the frequency of the different contributions are selected via the controls. There are sets of controls for six sine functions, plus the additional controls for the three characteristic functions: rectangle (strain softening), triangle (strain hardening), and sawtooth (shear bands). On the upper right is the window where the time domain data of either the simulation or the measured and the reconstructed signal is published. In the window below the subtraction of the two signals, the measured and the reconstructed signal, is shown. With the help of the controls, 'Overall Offset', 'Overall Phase', and 'Overall Amplitude' this deviation can be reduced. This will not influence the Fourier components of the reconstructed signal, and will just move the reconstructed signal on top of the measured signal for a better comparison. The indicator 'Sum' is a help to reduce the difference of the two signals. It is the sum of the absolute values of all data points, that are created by the subtraction of the measured and the reconstructed signal. For an optimal overlay this value has to be minimised. In the two windows below the frequency domain signal, the magnitude and the phase spectra, is shown. The resulting data can be saved via pressing the 'SAVE' button and an file can be selected after pressing 'OPEN NEW FILE'. The data that is saved to the file is the time domain data of the reconstructed or simulated signal plus the time domain signal after the Fourier-analysis in one file, and the values for the magnitudes and phases of the higher harmonics in the other file.

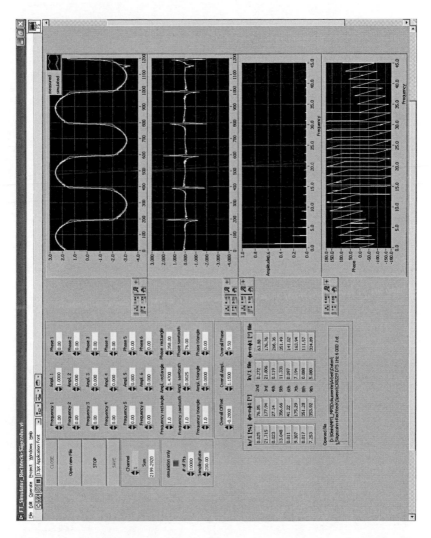

FIG. B.3: Panel of the virtual instrument 'Reconstruct Time Signal'. This is the interface for the analysis of time domain data with respect to the superposition of four characteristic functions: sinusoidal, rectangular, triangular and a sawtooth response.

B.3   VI Birefringence-Dichroism

The measurement software for the Rheo-optics is presented within this appendix. With this software the optical data is acquired from the two analog inputs, then treated to extract the optical observables of birefringence $\Delta n'$, dichroism $\Delta n''$ and the orientation angle θ. The data is then plotted together with the mechanical data and finally stored in data files.

B.3.1   VI Birefringence-Dichroism Main

The VI Birefringence-Dichroism is written to measure the birefringence or the dichroism and the orientation angle with the optical analysis module (OAM) from Rheometrics Scientific. At the left side of the front panel are the buttons, controls

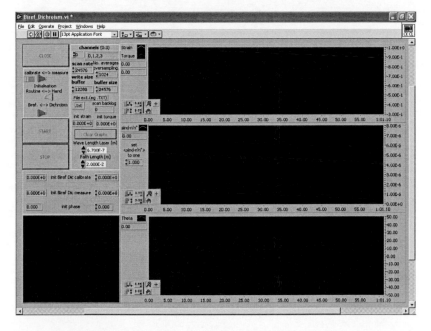

FIG. B.4: Panel of the virtual instrument 'BirefringenceDichroism'. This is the main interface for the rheo-optical measurement.

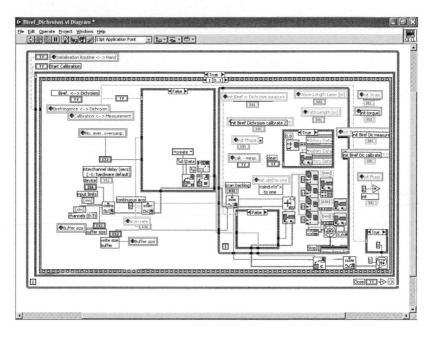

FIG. B.5: Diagram of the virtual instrument 'BirefringenceDichroism'.

and indicators. On the right side three displays, the upper most showing the mechanical data (strain and torque), the second showing the $sin\delta$ or the birefringence/dichroism and the bottom one shows the orientation angle θ. A forth display on the left bottom side shows the actual birefringence/dichroism vectors. Prior any calibration or measurement the 'path length' of the sample beam, the 'laser wavelength', and the 'number of averages oversampling/points for FT' are selected. Furthermore one has to switch the 'Biref. $\leftrightarrow$ Dichroism' button to the desired quantity. Note, that birefringence is only measurable when dichroism is negligible. First the calibration procedure is started with the preselected 'calibrate $\leftrightarrow$ measure' button. At this point it can be selected, if the initialisation procedure is performed automatically or not. If 'calibration by routine' is selected the initialisation window opens (see chapter B.3.2). After the initialisation the calibration values are visible in the corresponding 'init'-displays ('init strain', 'init torque', 'init Biref. Dic. calibrate', 'init Biref. Dic. measure' and 'init phase'). While making the calibration manually one has to put these values to zero with the above mentioned controls. After the transfer of the latter three values to the init-controls the calibration is checked. For the phase a linear polariser is placed in the sample beam and the value displayed in the graph should be at $\pm$ 45 °, which is equal to the maximum reachable signal. If not they can be readjusted with the 'init phase'-control. The $sin\delta$ value is checked via placing a $\frac{4}{\lambda}$ - plate in the beam. In our case the value should be 0.94 which is calculated from the thickness of the $\frac{4}{\lambda}$ - plate used here. The thickness of a $\frac{4}{\lambda}$ - plate depends on the wave length of the light λ for that is was designed. If this value does not fit a fine tuning is possible by using the 'set $\langle sind - n'n'' \rangle$ to one'-control. Whenever it is necessary or desired it is possible to clear the graph windows by pressing the 'CLEAR GRAPH' button. After calibration the button 'calibrate $\leftrightarrow$ measure' is switched to measurement and by pressing the 'Start' button the measurement is started. The measurement runs until the 'STOP' button is pressed. The data is automatically saved to a file. By pressing the 'CLOSE' button the software is stopped.

B.3.2   VI Birefringence-Dichroism Initialisation

After the start of the calibration routine within 'Birefringence-Dichroism' software, the initialisation routine is called. The focus of the window automatically shifts to the 'Initialisation Panel'. Here the averaged values of strain, torque, birefringence/dichroism, and orientation angle are taken over a specific time span. The duration of the averaging in this procedure is chosen by the user in the 'init panel' by the time control Fig. B.6. The default value is 20 s. By pressing the button 'Determine Initialisation Values' the averages are taken Fig. B.4 and after the given time the window will close and the focus of the window goes to the main panel B.3.1.

B.3.3   VI Birefringence-Dichroism Core

In the VI Birefringence-Dichroism Core (see Fig. B.8) the calculation, and the important correction of the input signal of birefringence/dichroism and orientation angle are computed. Figure B.7 displays the diagram of this program. On the left side of the diagram the data of the four channels is entered into a loop. The two upper channels (strain, torque) are averaged, while the signals of the other

FIG. B.6: Initialisation Panel with the time control.

two channels (reference and sample) are on-the-fly Fourier transformed. The FFT performed here works with 2^n data points. Doing so the fluctuation of the spinner signal is online monitored. Then the corrections from the initialisation and calibration routine are implemented and finally the birefringence/dichroism and the orientation angle values are calculated by the following equations:

$$I_S = I_{0,S}(1 + sin\delta' cos(4\omega_1 t(\theta - \frac{\pi}{4}))), \tag{B.1}$$

$$I_R = I_{0,R}(1 + cos(4\omega_1 t)), \tag{B.2}$$

with

$$4\omega_1/2\pi = 1600Hz, \tag{B.3}$$

where:

I_S: Signal intensity at the sample detector,

I_R: Signal intensity at the reference detector,

$I_{0,S}$: DC signal intensity at the sample detector,

$I_{0,R}$: DC signal intensity at the reference detector.

The birefringence is given by:

$$\delta' = \frac{2\pi\Delta n'd}{\lambda} = \frac{\pi}{2}, \tag{B.4}$$

is called the retardation δ', d the optical path length, λ the wavelength of the light source and $\Delta n' = n'_x - n'_y$ the birefringence. The orientation angle is calculated by:

$$tan2\theta = \frac{\Delta n'}{\Delta n''} = \frac{n'}{n''}. \tag{B.5}$$

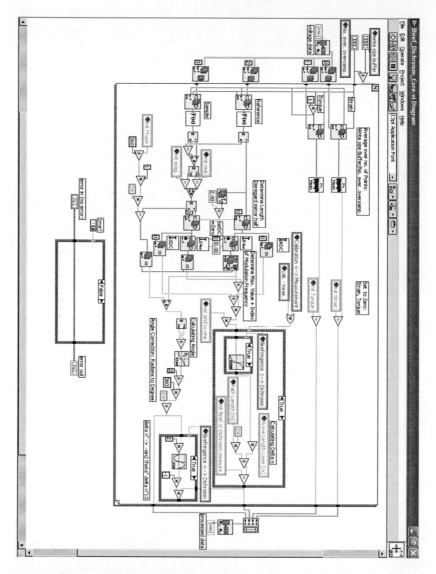

FIG. B.7: The Diagram of the VI 'BirefringenceDichroismCore'. Here the correction and the calculation of birefringence $\Delta n'$, dichroism $\Delta n''$ and orientation angle θ is performed.

FIG. B.8: The panel of the Sub-VI 'BirefringenceDichroismCore'. Input and Output Controls are visible.

Appendix C

Chemicals

CHEMICAL	ABBREVIATION	ORIGIN	PURITY [%]
water	H_2O	Milli-Q	$R = 18, 2\frac{M\Omega}{cm}$
n-butyl methacrylate	BMA	Fluka	≥ 99
acrylic acid	AA	Fluka	≥ 99
methyl methacrylate	MMA	Fluka	≥ 99
allyl methacrylate	AMA	Fluka	≥ 99
styrene	S	Fluka	$\geq 99,5$
sodium styrene sulfonate	NaSS	Aldrich	$\geq 99,5$
sodium dodecyl sulfate	SDS	Fluka	≥ 99
ammonium peroxodisulfate	APS	Fluka	≥ 99
potassium peroxodisulfate	KPS	Aldrich	$\geq 99,99$
azo-*bis*-isobutyrylnitrile	AIBN	Fluka	≥ 99
hexadecane	HD	Fluka	≥ 99
tris (hydroxy methyl)-amino methane chloride	TRIS-Cl	Fluka	≥ 99
di-*sec*-octyl phthalate	DOP	Fluka	$\geq 99,5$

Table C.1: origin and purity of the used chemicals

Appendix D

Synthesis

In this appendix the synthesis for the two basic routes, used for the synthesis of the polymer dispersions are presented. The first one is a semi-continuous emulsion polymerisation synthesis in two steps, consisting of a step for the synthesis of the seed and a second step that is characterised by a permanent inflow of reaction ingredients under starved feed conditions. After the initial step of a nucleation phase, also known as seed, a second step is performed. In this second step the reaction continuous during a permanent flow of the reactive medium. This inflow is realised via two motor driven syringes, which contain the reactive media. The first syringe contains monomer, SDS, acrylic acid and some water in a pre-emulsion. The second syringe contains additional water and the initiator. The second type of synthesis is done in one step. Here small droplets containing all necessary chemical are formed in which the reaction takes place. This method is called mini-emulsion polymerisation. First two mixtures are made in separate beakers. The first contains water, surfactant initiator, and the second containing monomer, hydrophobe, acrylic acid and co-monomer. After mixing these two beakers and stirring it for one hour, the solution is spiffed for 2 minutes at an amplitude of 89 %, which is the maximal usable amplitude for this sonifier tip. The reactive medium is cooled during the sonication. Afterwards the mixture is taken to the reactor and the synthesis is performed. The names of the samples are made up of two parts. The first, like CKA or CKB, denotes the first or the second lab journal and a second part denotes the number of the synthesis. Only the synthesis of the dispersions examined in this work are presented here.

D.1 Synthesis of semi-continuous emulsion polymerisation

POLYMER DISPERSION	CKB109	CKB117
operation mode	batch seed	semi-continuously
temperature [° C]	85	80
stirrer speed [rpm]	450	140
operating time [h]	7	5 for feed 1, 2
polymerisation time after reaction [h]	5	3
amount H_2O [g]	160	11
amount of seed [g]	-	16
amount monomer [g]	10 S	37 S
amount co-monomer [g]	-	0.7 AA
surfactant [g]	1 SDS	3.5 SDS
initiator [g]	0.3 KPS	0.1 KPS
prime [g]	160 H_2O	10 H_2O
feed 1	-	H_2O + SDS + S
feed 2	-	H_2O + KPS
solid content [%]	-	56.5
pH	3	8
particle diameter [nm]	55.2	99.7
polydispersity index	0.01	0.07

Table D.1: Dispersion CKB109 (seed), CKB117 (semi-continuous polymerisation).

POLYMER DISPERSION	CKB99	CKB103
operation mode	batch seed	semi-continuously
temperature [° C]	85	85
stirrer speed [rpm]	400	150
operating time [h]	7	5 for feed 1, 2
polymerisation time after reaction [h]	5	3
amount H_2O [g]	160	13
amount of seed [g]	-	16
amount monomer [g]	10 S	37 S
amount co-monomer [g]	-	0.7 AA
surfactant [g]	1 SDS	5.8 SDS
initiator [g]	0.3 KPS	0.1 KPS
prime [g]	160 H_2O	10 H_2O
feed 1	-	H_2O + SDS + S
feed 2	-	H_2O + KPS
solid content [%]	13	62.5
pH	-	10
particle diameter [nm]	-	80
polydispersity index	-	0.25

Table D.2: Dispersion CKB99 (seed), CKB103 (semi-continuous polymerisation).

POLYMER DISPERSION	CKB169	CKB171
operation mode	batch seed	semi-continuously
temperature [° C]	70	80
stirrer speed [rpm]	350	140
operating time [h]	0	5 for feed 1, 2
polymerisation time after reaction [h]	5	3
amount of seed [g]	-	20
amount H_2O [g]	200	15
amount monomer [g]	18.3	35 S
amount co-monomer [g]	0.6 MMA	0.7 AA
surfactant [g]	0.5 NaSS	4 SDS
initiator [g]	0.2 KPS	0.2 KPS
prime [g]	200 H_2O	20 seed
feed 1	-	H_2O + SDS + S
feed 2	-	H_2O + KPS
solid content [%]	-	55
pH	-	8
particle diameter [nm]	-	150
polydispersity index	-	0.3

Table D.3: Dispersion CKB169 (seed), CKB171 (semi-continuous polymerisation).

D.2 Synthesis of mini-emulsion polymerisation

POLYMER DISPERSION	CKB222	CKB232
operation mode	batch	batch
temperature [° C]	72	72
stirrer speed [rpm]	360	360
polymerisation time [h]	12	12
amount H_2O [g]	50	50
amount monomer [g]	15 S + 15 BMA	15 S + 15 BMA
surfactant [g]	0.4 SDS	0.3 SDS
initiator [g]	0.1 AIBN	0.1 AIBN
organic stabiliser [g]	0.5 HD	0.5 HD
prime	H_2O, SDS, AIBN	see CKB222
prime monomer	S + BMA	S + BMA
solid content [%]	32.7	35.3
yield [%]	96.7	99.1
pH	8	8
particle diameter [nm]	133.2	137
polydispersity index	0.01	0.02

Table D.4: Dispersion CKB222 and CKB232 (mini-emulsion polymerisation).

POLYMER DISPERSION	CKB235	CKB229
operation mode	batch	batch
temperature [° C]	72	72
stirrer speed [rpm]	360	360
polymerisation time [h]	12	12
amount H_2O [g]	50	50
amount monomer [g]	15 S + 15 BMA	30 S
co-monomer [g]	0.4 AA	0.4 AA
surfactant [g]	0.4 SDS	5.4 SDS
initiator [g]	0.1 V59	0.1 AIBN
organic stabiliser [g]	0.5 HD	0.5 HD
prime	H_2O, SDS, KPS	H_2O, SDS, KPS
prime monomer	S + BMA	S
solid content [%]	36.8	35.4
pH	8	8
particle diameter [nm]	68.3	69.7
polydispersity index	0.06	0.04

Table D.5: Dispersion CKB235 and CKB229 (mini-emulsion polymerisation).

Bibliography

[Ackerson 88]   B.J. Ackerson, P.N. Pusey. *Phys. Rev. Lett.* **61**, 1033 (1988).

[Amelar 91]   S. Amelar, C.E. Eastman, R.L. Morris, M.A. Smeltzly, T.P. Lodge, E.D. von Meerwall. *Macromolecules* **24**, 3505 (1991).

[Arendt 98]   O. Arendt, W.M. Kulicke. *Angew. Macromol. Chem.* **257**, 77 (1998).

[Arrhenius 17]   Z. Arrhenius. *Biochem. J.* **11**, 112 (1917).

[Ball 80]   R. Ball, P. Richmond. *J. Phys. Chem. Liquids* **99**, 9 (1980).

[Batchelor 71]   G.K. Batchelor. *J. Fluid. Mech.* **46**, 813 (1971).

[Bernal 60]   J.D. Bernal. *Nature* **185**, 68 (1960).

[Berne 76]   B.J. Berne, R. Pecora. Dynamic Light Scattering. Wiley, New York (1976).

[Bindell 92]   J.B. Bindell. Encyclopedia of Materials Characterization: Surfaces, Interfaces, Thin Films. In R. C. Brundle, C.A. Evans, S. Wilson (eds.), Applications of CAE in Extrusion and Other Continuous Processes. Butterworth-Heinemann, Greenwich (1992).

[Bitsanis 90]   I. Bitsanis, H.T. Davis, M. Tirrell. *Macromolecules* **23**, 1157 (1990).

[Böhnlein-Ma 92] J. Böhnlein-Mauss, W. Sigmund, G. Wegner, H.W. Meyer, F. Hessel, K. Seitz, A. Roosen. *Adv. Mater.* **4**, 73 (1992).

[Bossis 89]   G. Bossis, J.F. Brady. *J. Chem. Phys.* **91**, 1866 (1989).

[Bracewell 86]   R.N. Bracewell. The Fourier Transform And Its Application. McGraw-Hill, New York (1986).

[Brown 93]   W. Brown. Dynamic Light Scattering. Oxford Science Publications, Oxford (1993).

[Butler 95]   J.H. Butler, D.C. Joy, G.F. Bradley, S.J. Krause. *Polymer* **36**, 1790 (1995).

[Casey 94]   B.S. Casey, B.R. Morrison, I.A. Maxwell, R.G. Gilbert. *J. Polym. Sci.* **32**, 605 (1994).

[Claridge 99]   T.D.W. Claridge. High-Resolution NMR Techniques in Organic Chemistry. Pergamon, Amsterdam (1999).

[Clasen 01]   C. Clasen, W.M. Kulicke. *Rheol. Acta* **40**, 74 (2001).

[Colby 90]   R.H. Colby, M. Rubinstein. *Macromolecules* **23**, 2753 (1990).

[Collyer 98]   A.A. Collyer, D.W. Clegg. Rheological Measurement. Chapman and Hall, London (1998).

[Cooley 65]   J.W. Cooley, J.W. Tuckey. *Math. Comput.* **19**, 297 (1965).

[Couette 90]   M.M. Couette. *Am. Chem. Phys. Ser. VI* **21**, 433 (1890).

[Craciun 03]   L. Craciun, J.P. Carreau, M.-C. Heuzey, T.G.M. van de Ven, M. Moan. *Rheol. Acta* **42**, 410 (2003).

[Dames 01]   B. Dames, N. Willenbacher, R.M. Bradley. *Rheol. Acta* **40**, 434 (2001).

[Davis 78]   W.M. Davis, C.W. Macosko. *J. Rheol.* **22**, 53 (1978).

[Day 88]   L.A. Day, C.J. Marzec, S.A. Reisberg, A. Casadevall. *Ann. Rev. Biophys. Chem.* **17**, 509 (1988).

[deKruif 85]   C.G. deKruif, E.M.F. vanIersel, A. Vrij, Russel W.B. *J. Chem. Phys.* **83**, 4717 (1985).

[Derjaguin 41]   B.V. Derjaguin, L. Landau. *Acta Physicochim URSS* **10**, 25 (1941).

[Dhont 03]   J.K.G. Dhont, W.J. Briels. *Colloids and Surfaces A* **213**, 131 (2003).

[Distler 99]   D. Distler. Wäsrige Polymerdispersionen. Wiley VCH, Weinheim (1999).

[Dodge 71]   J.S. Dodge, I.M. Krieger. *Trans. Soc. Rheol.* **15**, 589 (1971).

[Doi 78]   M. Doi, S.F. Edwards. *J. Chem. Soc. Faraday Trans. II* **74**, 1789 (1978).

[Doi 81]   M. Doi. *J. Pol. Sci. Faraday Trans. II* **19**, 229 (1981).

[Doi 86]   M. Doi, S.F. Edwards. The theory of polymer dynamics, Volume 73 (*International series of monographs on physics*). Clarendon Press, Oxford, 2nd. edition (1986).

[Dörfler 02]   H.D. Dörfler. Grenzflächen und kolloid-disperse Systeme. Physik und Chemie. Springer, Berlin, 1st. edition (2002).

[Dusschoten 01]   D. van Dusschoten, M. Wilhelm. *Rheol. Acta* **40**, 395 (2001).

[Einstein 06]   A. Einstein. *Ann. Phys.* **19**, 289 (1906).

[Einstein 11]   A. Einstein. *Ann. Phys.* **34**, 591 (1911).

[El-Aasser 90]   M.S. El-Aasser. An Introduction to Polymer Colloids. In F. Candau, R.H. Ottewill (eds.), Observation, Prediction and Simulation of Phase Transitions in Complex Fluids. Kluweer Academic Publishers, Dordrecht (1990).

[El-Aasser 97]   M.S. El-Aasser, P.A. Lovell. Miniemulsion Polymerization. In M.R. Baus, J.P. L.F. Ryckaert (eds.), Emulsion Polymerization and emulsion polymers. Wiley, Chichester (1997).

[Fearn 99]   T. Fearn. *Spectr. Europe* **11**, 24 (1999).

[Fraden 95]   S. Fraden. Phase transition in colloidal suspensions of virus particles. In M. Rull Baus, J.P. L.F. Ryckaert (eds.), Observation, Prediction and Simulation of Phase Transitions in Complex Fluids, Volume 460. Kluweer Academic Publishers, Dordrecht, 1st. edition (1995).

[Fuller 90]   G.G. Fuller. *Annu. Rev. Fluid Mech.* **22**, 387 (1990).

[Fuller 95]   G.G. Fuller. Optical Rheometry of complex fluids. Oxford University Press, New York (1995).

[Gedde 95]   U. Gedde. Polymer Physics. Chapman and Hall, London (1995).

[Giacomin 98]   A.J. Giacomin, J.M. Dealy. Large-Amplitude Oscillatory Shear. In A.A. Collyer, D.W. Clegg (eds.), Rheological Measurements. Chapman and Hall, London (1998).

[Gilbert 95]   R.G. Gilbert. Emulsion Polymerization, a Mechanistic Approach. Academic Press, London (1995).

[Goldstein 92]   J.I. Goldstein, A.D. Romig, D.E. Newbury, C.E. Lyman, P. Echlin, C. Fiori, D.C. Joy, E. Lifshin. Scanning Electron Microscopy and X-Ray Microanalysis. Plenum, New York (1992).

[Graham 95]   M.D. Graham. J. Rheol. 39, 697 (1995).

[Hansen 78]   F.K. Hansen, J Ugelstad. Polym. Sci. 16, 1953 (1978).

[Harkins 47]   W.D. Harkins. J. Am. Chem. Soc. 69, 1428 (1947).

[Harkins 50]   W.D. Harkins. J. Polym. Sci. 5, 217 (1950).

[Helfand 82]   E. Helfand, D.S. Pearson. J. Polym. Sci. Polym. Phys. 20, 1249 (1982).

[Heymann 01]   L. Heymann, S. Peukert, N. Aksel. J. Rheol. 46, 93 (2001).

[Higgins 76]   R.J. Higgins. Am. J. Phys. 44, 766 (1976).

[Hilliou 02]   L. Hilliou, D. Vlassopoulos. Ind. Eng. Chem. Res. 41, 6246 (2002).

[Hofmann 09]   F. Hofmann, K. Dellbruck, K. Gottlob. Dt. Patent DRP 250, 690 (1909).

[Homans 89]   S.W. Homans. A Dictionary of Concepts in NMR. Clarendon Press, Oxford (1989).

[Honerkamp 94]   J. Honerkamp. Stochastic Dynamical Systems: Concepts, Numerical Methods, Data Analysis. VCH, New York (1994).

[Hoy 79]   K.L. Hoy. Techn. 51, 27 (1979).

[Hunter 88]   R.J. Hunter. Foundations of colloidal science. Clarendon Press, Oxford (1988).

[Hyun 02]   K. Hyun, S.H. Kim, K.H. Ahn, S.J. Lee. J. Non-Newt. Fluid Mech. 107, 51 (2002).

[Jaksch 95]   H. Jaksch, J.P. Martin. Fresenius J. Anal. Chem. 353, 378 (1995).

[Janeschitz-Kr 83] H. Janeschitz-Kriegl. Polymer melt rheology and flow birefringence. Springer, Berlin (1983).

[Janssen 93]   R.Q.F. Janssen, G.J.D. Derks, A.M. van Herk, A.L. German. Encapsulation. In D.R. Karsa, R.A. Stephenson (eds.), Encapsulation and Controlled Release. Royal Society of Chemistry, Cambridge (1993).

[Joy 96]   D.C. Joy, C.S. Joy. Micron. 27, 247 (1996).

[Kallus 01]   S. Kallus, N. Willenbacher, D. Kirsch, D. Distler, T. Neidhöfer, M. Wilhelm, H.W. Spiess. *Rheol. Acta* **40**, 552 (2001).

[Kerker 69]   M. Kerker. The Scattering of Light and Other Electromagnetic Radiation. Academic Press, New York (1969).

[Keunings 04]   R. Keunings, K. Atalik. *J. Non-Newtonian Fluid Mech.* **in press** (2004).

[Krieger 59]   I.M. Krieger, T.J. Dougherty. *Trans. Soc. Rheol.* **3**, 137 (1959).

[Krieger 63]   I.M. Krieger. *Trans. Soc. Rheol.* **7**, 101 (1963).

[Krieger 73]   I.M. Krieger, T.F. Niu. *Rheol. Acta* **12**, 567 (1973).

[Kulicke 98]   W.M. Kulicke, U. Reinhardt, O. Arendt. *Macromol. Rapid Commun.* **38**, 219 (1998).

[Kumaraswamy 99] G. Kumaraswamy, A.M. Issaian, J.A. Kornfield. *Macromolecules* **32**, 7537 (1999).

[Kuzuu 83]   N. Kuzuu, M. Doi. *J. Chem. Soc. Japan* **52**, 3486 (1983).

[LabVIEW 98]   Data Acquisition Basics Manual LabVIEW, January 1998 Edition. National Instruments (1998).

[Lambla 85]   M. Lambla, B. Schlund, E. Lazarus, T. Pith. *Macromol. Chem.* **10**, 463 (1985).

[Landfester 03] K. Landfester, F.J. Schork, V.A. Kusuma. *C. R. Chimie* **6**, 1337 (2003).

[Lange 89]   F.F. Lange. *J. Am. Ceram. Soc.* **72**, 3 (1989).

[Larson 99]   R.G. Larson. The Structure and Rheology of Complex Fluids. Oxford University Press, Oxford (1999).

[Laun 92]   H.M. Laun, R. Bung, S. Hess, W. Loose, O. Hess, K. Hahn, T. Hädicke, R. Hingmann, F. Schmidt, P. Lindner. *J. Rheol.* **36**, 743 (1992).

[Lechner 96]   M.D. Lechner, K. Gehrke, E.H. Nordmeier. Makromolekulare Chemie. Birkhäser Verlag, Basel, 2st. edition (1996).

[Li 00]   Y. Li, C.W. Park. *Adv. Colloid Interface Sci.* **87**, 1 (2000).

[Lide 96]   D.R. Lide. Handbook of Chemistry and Physics. CRC, New York (1996).

[Linden 03]   E. van der Linden, L. Sagis, P. Venema. *Curr. Opin. Colloid In.* **8**, 349 (2003).

[Liu 75]   S.T. Liu, G.H. Nancollas. *J. Colloid Interface Sci.* **52**, 582 (1975).

[Lodge 94]   T.P. Lodge, C.W. Macosko. Rheo-optics: Flow birefringence. VCH publishers, Inc., New York (1994).

[Macosko 94]   C.W. Macosko. Rheology: Principles, Measurements, and Applications. VCH publishers, Inc., New York (1994).

[Marvin 75]   E.J. Marvin D.A. Wachtel. *Nature* **253**, 19 (1975).

[Matsumoto 73]   T. Matsumoto, Y. Segawa, Y. Warashina, S. Onogi. *Trans. Soc. Rheol.* **17**, 47 (1973).

[Meeker 97]   S.P. Meeker, W.C.K. Poon, P.N. Pusey. *Phsy. Rev. E* **55**, 5718 (1997).

[Mewis 89]   J. Mewis, W.J. Frith, T.A. Strivens, W.B. Russel. *AIChE J.* **35**, 415 (1989).

[Mewis 97]   J. Mewis, M. Mortier, J. Vermant, P. Moldenaers. *Macromolecules* **30**, 1323 (1997).

[Mori 82]   Y. Mori, N. Ookubo, R. Hayakawa, Y. Wada. *J. Polym. Sci.* **20**, 2111 (1982).

[Morrison 94]   B.R. Morrison, B.S. Casey, L. Laclk, G.L. Leslie, D.F. Sangster, R.G. Gilbert, D.H. Napper. *J. Polym. Sci.* **32**, 631 (1994).

[Neidhöfer 01]   T. Neidhöfer, M. Wilhelm, H.W. Spiess. *Appl. Rheol.* **11**, 126 (2001).

[Neidhöfer 03a] T. Neidhöfer. *Fourier-transform rheology on anionically synthesised polymer melts and solutions of various topology.* Ph.d. thesis, Johannes-Gutenberg-Universität, Mainz, Germany (2003).

[Neidhöfer 03b] T. Neidhöfer, M. Wilhelm, B. Debbaut. *J. Rheol.* **47**, 1351 (2003).

[Newman 77]   J. Newman, H.L. Swinney, L.A. Day. *J. Mol. Biol.* **116**, 593 (1977).

[Öner 98]   M Öner, J. Norwig, H.W. Meyer, G. Wegner. *Chem. Mater.* **10**, 460 (1998).

[Onogi 70]   S. Onogi, T. Masuda, K. Kitagawa. *Macromolecules* **3**, 109 (1970).

[Pearl 92]   A.S. Pearl. *Ceram. Bull.* **71**, 821 (1992).

[Pearson 82]   D.S. Pearson, W.E. Rochefort. *J. Polym. Sci. Polym. Phys.* **20**, 83 (1982).

[Pecora 85]   R. Pecora. Dynamic Light Scattering. Plenum, New York (1985).

[Peterlin 76]   A. Peterlin. *Annu. Rev. Fluid Mech.* **8**, 35 (1976).

[Phung 96]   T. Phung, J.F. Bradley, G. Bossis. *J. Fluid. Mech.* **313**, 181 (1996).

[Piirma 82]   I. Piirma. Emulsion Polymerisation. Academic Press, London (1982).

[Pipkin 72]   A.C. Pipkin. Lectures on viscoelastic theory. Springer-Verlag, New York (1972).

[Provencher 82a] S.W. Provencher. *Computer Physics Communication* **27**, 213 (1982).

[Provencher 82b] S.W. Provencher. *Computer Physics Communication* **27**, 229 (1982).

[Quemada 78]   D. Quemada. *Rheol. Acta* **17**, 632 (1978).

[Ramirez 85]   R.W. Ramirez. The FFT Fundamentals and Concepts. Engelwood Cliffs, Prentice-Hall (1985).

[Regitz 99]   M. Regitz, J. Falbe. Römpp Lexikon Chemie. Thieme, Stuttgart (1999).

[Reimers 96]   M.J. Reimers, J.M. Dealy. *J. Rheol.* **40**, 167 (1996).

[Reimers 98]   M.J. Reimers, J.M. Dealy. *J. Rheol.* **42**, 527 (1998).

[Rice 90]   R.W. Rice. *AIChE J.* **36**, 481 (1990).

[Rueb 98]   C.J. Rueb, C.F. Zukoski. *J. Rheol.* **42**, 6 (1998).

[Sagis 01]   L.M.C. Sagis, E. van der Linden, M. Ramaekers. *Phys. Rev. E* **63**, 051504 (2001).

[Sawyer 96]   D.T. Sawyer, D.T. Grubb. Polymer Microscopy. Chapman and Hall, London (1996).

[Schmidt-Rohr 94] K. Schmidt-Rohr, H.W. Spiess. Multidimensional Solid-State NMR and Polymers. Academic Press, London (1994).

[Schramm 90]   G. Schramm. An Introduction to Dynamic Light Scattering by Macromolecules. Academic Press, London (1990).

[Schramm 95]   G. Schramm. Einführung in die Rheologie und Rheometrie. Haake, Karlsruhe (1995).

[Sim 03]   H.G. Sim, K.H. Ahn, S.J. Lee. *J. Rheol.* **47**, 879 (2003).

[Simon 93]   C. Simon. Stabilization of Aqueous Powder Suspensions in the Processing of Ceramic Materials. In B. Dobais (eds.), Coagulation and Flocculation. Marcel Dekker, New York (1993).

[Skoog 92]   D.A. Skoog, J.J. Leary. Principles of Instrumental Analysis. Harcourt Brace College Publishers, Orlando, 4th. edition (1992).

[Smith 48]   V.W. Smith, R.H. Ewart. *J. Chem. Phys.* **16**, 592 (1948).

[Solomon 93]   E.I. Solomon, P.M. Jones, J.A. May. *J. A. Chem. Rev.* **93**, 2623 (1993).

[Song 89]   Z. Song, G.W. Poehlein. *J. Coll. Interf. Sci.* **128**, 501 (1989).

[Tariq 98]   S. Tariq, A.J. Giacomin, S. Gunasekarab. *Biorheol.* **35**, 171 (1998).

[Taylor 23]   G.I. Taylor. *Phil. Trans.* **A223**, 289 (1923).

[Teraoka 89]   I. Teraoka, R. Hayakawa. *J. Chem. Phys.* **91**, 2643 (1989).

[Vanderhoff 85] J. W. Vanderhoff. *J. Polym. Sci.* **72**, 161 (1985).

[vanWazer 63]   J.R. vanWazer, J.W. Lyons, K.Y. Lim, R.E. Colwell. Viscosity and Flow Measurement. Wiley, New York (1963).

[Ven 90]   T.G.M. Van de Ven. The Application of Flow Birefringence to Rheological Studies of Polymer Melts. In R.H. Canadau F. Ottewill (eds.), Scientific Methods for the Study of Polymer Colloids and Their Applications, p. 247. Kluwer Academic (1990).

[Verwey 48]   J.W. Verwey, J.T.G. Overbeek. Theory of the Stability of Lyophobic Colloid. Elsevier Publ Comp, Amsterdam (1948).

[Vezie 95]   D.L. Vezie, E.L. Thomas, W.W. Adams. *Polymer* **36**, 1761 (1995).

[Visscher 94]   P.B. Visscher, D.M. Heyes. *J. Phys. Chem.* **101**, 6096 (1994).

[Wagner 98]   N.J. Wagner. *Curr. Opin. Colloid Interface Sci.* **3**, 391 (1998).

[Wiese 92]   H. Wiese. *GIT Fachz. Lab.* **36**, 3385 (1992).

[Wilhelm 98]   M. Wilhelm, D. Maring, H.W. Spiess. *Rheol. Acta* **37**, 399 (1998).

[Wilhelm 99]   M. Wilhelm, P. Reinheimer, M. Ortseifer. *Rheol. Acta* **38**, 349 (1999).

[Wilhelm 00]   M. Wilhelm, P. Reinheimer, M. Ortseifer, T. Neidhöfer, H.W. Spiess. *Rheol. Acta* **39**, 241 (2000).

[Wilhelm 02]   M. Wilhelm. *Macromol. Mater. Eng.* **287**, 83 (2002).

[Xia 00]   Y. Xia, B. Gates, Y. Yin, Y. Lu. *Adv. Mater.* **12**, 693 (2000).

Acknowledgements

In erster Linie möchte ich mich bei Herrn *Prof. Dr. Hans Wolfgang Spiess* für die Möglichkeit bedanken, am MPIP meine Doktorarbeit zu erstellen. Die Freude, das Interesse, die Begeisterung und zahlreichen Anregungen, die er vor allem bei der Entwicklung der "Neuen Methode zur Analyse von nichtlinearen rheologischen Zeitdaten" zeigte, waren immer wieder ein Ansporn mit Begeisterung daran weiter zuarbeiten. Vielen Dank für die aufmunternden Worte: wie z.B.: "illuminating" und "Highlight".

Ich möchte besonders meinem Projektleiter *Prof. Dr. Manfred Wilhelm* für die Bereitstellung des interessanten Themas danken. Trotz Deiner oft starken Belastung durch Deine Habilitation, Deine Vorbereitung auf die neue Stelle und nicht zuletzt Deiner Familie, hast Du Dir immer Zeit für Besprechungen genommen und mit einer Unmenge an Ideen zur Lösung von Problemen beigetragen. Hervorheben möchte ich Deine Unterstützung bei der Planung und der Durchführung meiner unvergeßlichen Forschungsaufenthalte in den Niederlanden.

Vielen Dank an ... / I am deeply indebted to ...

Prof. Tadeusz Pakula for using his ARES rheometer including the Optical Analysis Module II.

Dr. Loïc Hilliou and *Dr. Dagmar van Dusschoten* for their expertise in the field of Rheology and Rheo-optics.

Prof. Eric van der Linden for the possibility to work in his group (Wageningen University, the Netherlands).

Dr. Paul Venema, Dr. Leonard Sagis, Dr. Henny Schaink and all the members from the food-physics-group in Wageningen for their cooperation and friendship.

215

Prof. Dr. Gerhard Wegner, Dr. Franziska Gröhn, Gerda Jentzsch for the cooperation in the project: "Mineralisation under shear".

Prof. Dr. Jan Dhont for providing the theory on rod-like particles and *Dr. Pavlik Lettinga* for supplying me with the FD-virus sample and fruitful discussions.

Deutsche Forschungsgemeinschaft für die finanzielle Unterstützung: DFG WI 1911/1.

The European Union for the grant:
Marie-Curie Trainingssite IMP HPMT-2000-00188.

BMBF-Verbundvorhaben für die finanzielle Unterstützung (Förderkennzeichen: 01RC0175):
"Wasser als Medium bei Herstellung, Verarbeitung und Anwendung von Kunststoffen"
im Förderschwerpunkt:
"Integrierter Umweltschutz in der Kunststoff- und Kautschukindustrie".

Allen Mitgliedern des AK Spiess.

Besonders möchte ich mich bei meiner Familie für die Unterstützung während der Doktorarbeit und natürlich darüber hinaus sehr herzlich bedanken.

List of Figures

List of Tables

Curriculum Vitae

Persönliche Daten

Name: Christopher Olaf Klein

Geburtsdaten: 21. Mai 1974 in Bad Kreuznach

Nationalität: deutsch

Familienstand: ledig

Schulausbildung

1980 - 1984   Grundschule an der Kleiststrasse, Bad Kreuznach, Rheinland-Pfalz

1984 - 1993   Lina-Hilger Gymnasium, Bad Kreuznach, Rheinland-Pfalz

Zivildienst

1993 - 1994   Zivildienst, Rettungssanitäter Malteser Hilfsdienst e.V.,

                      Bad Kreuznach, Rheinland-Pfalz

Studium

10/94 - 11/99   Universität Kaiserslautern Studiengang Chemie

11/99 - 09/00   Diplomarbeit Universität Kaiserslautern, unter der Betreuung
von Prof. R. Wortmann mit dem Titel:
"Elektrooptische Absorptionsmessungen zur Charakterisierung und
Optimierung von Chromophoren für photorefraktive Polymere"

10/00   Diplom in Chemie

10/00 - 12/04   Doktorarbeit am Max-Planck Institut für Polymerforschung, Mainz,
unter der Betreuung von Prof. H.W. Spiess und PD Dr. M. Wilhelm
mit dem Titel:
"Rheology and Fourier-Transform Rheology on water-based systems"

04/01 - 07/01   Forschungsaufenthalt in der Arbeitsgruppe von Prof. E. v. d. Linden,
Department of Food-Physics, University of Wageningen, Niederlande

05/03 - 08/03   Forschungsaufenthalt in der Arbeitsgruppe von Prof. E. v. d. Linden,
Department of Food-Physics, University of Wageningen, Niederlande